Plastic Injection Molding

Plastic Injection Molding

...manufacturing
process
fundamentals

By Douglas M. Bryce

Volume I: *Fundamentals of Injection Molding* series

Published by the
Society of Manufacturing Engineers
Dearborn, Michigan

Library of Congress Catalog Card Number: 96-067394
International Standard Book Number: 0-87263-472-8

Additional copies may be obtained by contacting:

Society of Manufacturing Engineers
Customer Service
One SME Drive
Dearborn, Michigan 48121
1-800-733-4763

SME staff who participated in producing this book:

Donald A. Peterson, Senior Editor
Dorothy M. Wylo, Production Assistant
Rosemary K. Csizmadia, Operations Administrator
Sandra J. Suggs, Editorial Assistant
Jerome T. Cook, Staff Photographer
Karen M. Wilhelm, Manager, Book Publishing
Cover design by Judy D. Munro, Manager, Graphic Services

Printed in the United States of America

Table of Contents

Chapter 3 - Parameters of the Molding Process

Chapter 5 - The Role of the Operator

Chapter 6 - Basics of Materials

Chapter 9 - Secondary Operations

Chapter 10 - Testing and Failure Analysis

List of Tables and Figures

Chapter 4

Chapter 5

Chapter 6

Chapter 7

Chapter 8

Chapter 9

Chapter 10

Chapter 11

Preface

This book (and accompanying volumes to follow) represents over three decades of my involvement in the plastics industry, most of which was spent in injection molding of thermoplastic materials. Through those years, it became apparent to me that most of the people in this industry had learned what they know by *doing*. Their skills were honed by making mistakes, learning from those mistakes, and plunging forward to discover other areas in which the learning process had to be repeated. While this method of attaining knowledge did work, I felt it would be better if novices to the industry could be made aware of the *basics* before being exposed to the trials and tribulations that accompany typical seat-of-the-pants accumulation of knowledge.

Before deciding to write a book on the subject, I researched the available literature and found it to be written, for the most part, for those already in the industry who had a working knowledge of the day-to-day routines involved with injection molding. So, I decided to write a source of basic, fundamental information on injection molding for those who are interested in getting a sound initial grasp of the subject and who also wish to have a reference tool of charts, diagrams, and data that can be used for years to come. I believe this goal is accomplished in the publication of this series.

I have structured the text as a guide to lead the reader through the entire injection molding process from its historic inception to the current state of the art. At the close of each chapter are questions pertinent to the material of that chapter. Answers to the questions appear at the end of the book.

ACKNOWLEDGMENTS

This work could not have been done without the cooperation and involvement of many people. I wish to take this opportunity to thank the following for their contributions to the achievement of this goal:

- *Society of Manufacturing Engineers* for having the foresight and courage to publish the book(s).
- *AEC, Incorporated,* for the use of photographs and information concerning various pieces of equipment manufactured by them.

- *Branson Ultrasonics, Incorporated,* for the use of photographs and information concerning various pieces of equipment manufactured by them.
- *GE Plastics* for the use of photographs and information concerning various plastics manufactured by them.
- *Perkin-Elmer Corporation* for the use of photographs and information concerning various pieces of equipment manufactured by them.
- *Texas Plastic Technologies* for allowing me the time and opportunity to write and research on the job.
- *United Silicone, Incorporated,* for the use of photographs and information concerning various pieces of equipment manufactured by them.

I am especially grateful to *my wife, my family, and my God* for the encouragement and support they delivered during the time it took me to write the book, and for putting up with my seemingly unlimited requests and tremendous mood swings during that period.

Please accept this book as it is intended, and I welcome you to the world of plastics. Here's hoping your involvement in the injection-molding industry will be as rewarding as mine has been.

Douglas M. Bryce
Georgetown, Texas 1996

Overview of the Industry

1

HOW IT ALL BEGAN

In 1868, an enterprising young gentleman by the name of John Wesley Hyatt developed a plastic material he called *celluloid* to enter in a contest created by a billiard ball manufacturer. The company was looking for a new material to substitute for ivory, which was becoming expensive and difficult to obtain. Celluloid was actually invented in 1851 by Alexander Parkes, but Hyatt improved it so that it could be processed into finished form. He created a celluloid billiard ball and won the contest's grand prize of $10,000, a rich sum in those days. Unfortunately, after the prize was won, some billiard balls exploded on impact during a demonstration (due to the instability and high flammability of celluloid) and further perfection was required before it could be used in commercial ventures. But the plastics industry was born and began to flourish when Hyatt and his brother Isaiah patented the first injection-molding machine in 1872. With this machine, the brothers were able to mold celluloid plastic. Over the next 40 to 50 years, others began to investigate this new process for manufacturing such items as collar stays, buttons, and hair combs. By 1920, the injection-molding industry had set its foundation, and it has been building ever since.

During the 1940s, the plastic injection-molding industry exploded with a bang (*not* because of the instability of celluloid) as World War II created a demand for inexpensive, mass-produced products. New materials were invented for the process on a regular basis and technical advances resulted in more and more successful applications.

EVOLUTION OF THE SCREW

The machine that the Hyatt brothers invented was primitive, but performed well for their purposes. It was simple in that it acted like a large hypodermic needle and contained a basic plunger to inject the plastic through a heated cylinder into a mold. In 1946, James Hendry began marketing his recently patented screw injection machine. This auger design replaced the conventional Hyatt plunger device and revolutionized the processing of plastics. Screw machines now account for approximately 95 percent of all injection machines.

The auger design of the screw creates a mixing action in material being readied for injection. The screw is inside the heating cylinder and, when activated, mixes the plastic well, creating a homogenized blend of material. This is especially useful when colors are being added or when *regrind* (recycled material) is being mixed with virgin material. After mixing, the screw stops turning and the entire screw pushes forward, acting like a plunger to inject material into the mold.

Another advantage of the screw is reduced energy requirements. As in a plunger machine, the cylinder that holds the plastic for injection has a series of electrical heater bands around the outside. When energized, these bands heat up and soften the plastic. However, the screw creates friction when it turns within the cylinder and thus generates additional heat. Therefore, the material is also heated from the inside, and less heat is required from the electrical heater bands to soften the plastic.

Although the screw machine is by far the most popular, there is still a place for the plunger-type machine. A plunger does not rotate. It simply pushes material ahead, then retracts for the next cycle. It, too, resides within a heated cylinder. Because there is no rotating, there is no shearing or mixing action. So, in a plunger machine, heat is provided solely by the external heater bands because there is no friction from the plunger as there is from a screw. If two different colored materials are placed in the heated cylinder, they are not blended together. The plunger simply injects the materials at the same time. If the two colors are, for instance, white and black, the resulting molded part will take on a marbled appearance with swirls of black and white. This may be the desired finish for certain products, such as lamp bases or furniture, and the plunger machine allows that finish to be molded into the product. Use of a screw machine would result in a single-color (gray) product being molded because the two colors would be well mixed prior to injection.

INDUSTRY EVOLUTION

From its birth in the late 1800s to the present time, the injection-molding industry has grown at a fast and steady rate. It has evolved from producing combs and buttons to molding products for varied industries, including automotive, medical, aerospace, consumer, toys, plumbing, packaging, and construction.

Table I-1 lists some of the important dates in the evolution of the injection-molding industry.

Table I-1. Evolution of Injection Molding

1868	John Wesley Hyatt injection-molds celluloid billiard balls.
1872	John and Isaiah Hyatt patent the injection-molding machine.
1937	Society of the Plastics Industry founded.
1938	Dow invents polystyrene (still one of the most popular materials).
1940	World War II creates large demand for plastic products.
1941	Society of Plastics Engineers founded.
1942	DME introduces stock mold base components.
1946	James Hendry builds first screw injection-molding machine.
1955	General Electric begins marketing polycarbonate.
1959	DuPont introduces acetal homopolymer.
1969	Plastics land on the moon.
1972	The first part-removal robot is installed on a molding machine.
1979	Plastic production surpasses steel production.
1980	Apple uses acrylonitrile-butadiene-styrene (ABS) in the Apple IIE computer.
1982	The JARVIK-7 plastic heart keeps Barney Clark alive.
1985	Japanese firm introduces all-electric molding machine.
1988	Recycling of plastic comes of age.
1990	Aluminum mold introduced for production molding.
1994	Cincinnati-Milacron sells first all-electric molding machine in the U.S.

A VISION OF TOMORROW

The future will see some major changes in the way injection-molding companies operate. In particular, changes will take place in four principal areas: processes, materials, molds (tooling), and business concepts.

Processes

Energy Efficiency

Energy-efficient machines will be developed to better utilize available resources. At present, injection-molding machines use vast amounts of electricity to heat the plastic, power the hydraulic pumps and motors, and control the temperature of the molds. The cost of this energy is steadily rising and the resources used to create the electricity are becoming more scarce and consequently more expensive. Thus, it is necessary to find ways to reduce the amount of energy required to produce products.

Some of the innovations being considered to reduce energy requirements include internally heated injection screws, insulated molds, and insulated heating cylinders. The combined use of these three items alone could result in energy savings of 60 percent or more when efficient models become available.

In addition, work is being done to develop all-electric machines. This concept takes the electricity now being used to power hydraulic systems and uses it directly to power electric motors instead. The motors then provide the motions normally provided by hydraulics, and the result is a more energy-efficient total system. At present, these systems are available only on small machines but as they become larger and less expensive, their popularity will increase, especially in clean-room environments and areas where noise must be reduced.

Desktop Manufacturing

The concept of "desktop manufacturing" (DM) has given rise to a brand new approach to injection-molding processes. DM can be defined in this case as molding products by using just a few cavities on high-volume equipment, small enough to fit in an area no larger than a desktop.

Although desktop manufacturing is already available for several types of production, the DM discussed here refers only to injection-molding processes. It is now possible to set up a large bank of small, benchtop injection-molding machines, each running only a one- or two-cavity mold, pumping out products much faster than the bulky, multicavity systems traditionally employed. With DM, a molder may elect to run several presses to make the same product, or only a few at a time, depending on immediate requirements. This allows more flexibility in the overall scheduling process. DM can also reduce manufacturing costs owing to less energy needed and faster cycles possible because of the smaller size of the equipment. Mold repairs can be made on one cavity while the rest of the cavities continue to run, unlike common injection processes today.

Desktop manufacturing is ideal for prototyping. In this situation, DM allows fast manufacturing of a few samples of a product, which then can be used for form, fit, and function evaluation prior to investing in production tooling. And DM is perfect for small-volume production where only a few hundred (or thousand) pieces are needed.

Molds

Tooling

Because of advances in molding machines and moldmaking equipment, there will be a trend to build both larger and smaller molds.

Molding machine manufacturers are building larger machines to accommodate product designs that were not possible in the past because of molding-machine size constraints. Products such as automotive fenders and wraparound bumpers will be molded on machines that are the size of small houses. The more this is done, the more it will drive manufacturers to build larger and larger machines. Of course, every machine must have a mold. So the molds will be built larger and larger to accommodate product design requirements.

Conversely, advances in materials and processing systems have led to the production of small parts that formerly were not candidates for injection molding. Products such as miniature electronic connectors and tiny medical valves are now being designed. The tolerance requirements and small size of these components require extremely accurate, sophisticated molding machines; these are being built now in sizes that will fit on the top of an ordinary desk. The machine tool manufacturers are striving to build even smaller machines as the demand increases. So, the molds for these machines are also smaller and smaller. There are molds now that can fit in the palm of a human hand, and the trend is toward even smaller molds to accommodate future product requirements.

Lead Times

Lead time is defined as the total amount of time required to obtain a product, from purchase order to finished item. In other words, the lead time for a mold extends from the moment a purchase order is received by the moldmaker to the moment the mold is delivered to the company ordering the mold. A typical lead time today ranges from 12 to 16 weeks for an average mold. (This does not include time to debug the mold or try it out in a production environment.)

A lead time of 12 to 16 weeks may not seem like much to those who have been in the business a while. In fact, it is a great improvement over what prevailed before. In earlier days, the lead time may have been 36 weeks or longer. Nonetheless, in today's competitive environment, lead times are critical because they dictate when a product can get to its market. The earlier the product can be introduced, the faster it can begin bringing in profits, and the quicker the company can begin investing those profits in new product development.

There are numerous ways to minimize lead times, and many are being pursued today. Certain of these promise to bring lead times down to unheard-of numbers:

- *Computer-generated data.* Computers allow moldmakers and product designers to work closely together even if they are in different cities, states, or countries. As computers become faster and more powerful, and as computer programs become more versatile, product designs can be generated and tested faster, and the same data can be used to make the molds. In fact, these things can be done as parallel efforts so the mold can be started before the product design is even finished.

- *Mold materials.* The common practice has been to use high-grade tool steels to build the cavities of the injection mold. However, new alloys and upgrades to present alloys are allowing molds to be built faster and weigh less. One material that will be used extensively is aluminum. Aluminum is now used primarily for prototype molds, but advances in the material alloys and acceptance by moldmakers has made it possible to use aluminum in many production mold cases. Beryllium copper, brass, soft steels, and even plastics such as epoxies are being used more and more in an effort to reduce lead times for making molds and thus get products into the marketplace faster.

The use of these materials and new computer equipment and programs can drastically reduce lead times. In some cases, lead times have been reduced from the average 12 to 16 weeks to only 7 days. And the technology will soon be available to bring it down to only a few hours.

Materials

Advances in plastic materials have been profound. In 1995, there were approximately 18,000 different materials available for molding. These are increasing at an average rate of 750 per year. The majority of available materials are alloys or blends of previously developed materials, and a product designer will probably be able to choose from them a material that provides exactly the right properties for a specific requirement. Again, computers come into play because the designer cannot possibly read through the property values of all those materials in a reasonable amount of time. The computer can do it in a few seconds and will list the materials that meet whatever criteria the designer requires. Then the designer can make a choice from only a few materials rather than 18,000. But even if an exact material match does not exist, the designer will be able to call on a *compounder* to put together a material that does exactly match the requirements. Of course, that adds another material to the list of 18,000, and that's how the list evolved in the first place.

Recycling

Recycling will continue to be a major issue in the future, as it is today. Consumer acceptance of products made of recycled materials will increase, making it more profitable for companies to develop such products. Technological advances will make it easy to separate discarded plastic products so they may be properly recycled. Advances in machinery and material additives will allow mingled plastics (that cannot be separated) to be used in products such as parking lot bumpers, picnic tables/benches, and wa-ter-sport products such as boating docks. Some of these products are available today, but with future advances, they can be sold at more competitive prices, and their availability and use will grow. Products will be designed with recycling in mind so they may be easily dismantled and identified for material content.

Business Concepts

Education and Training

As more and more companies downsize to become more efficient, em-ployees find that they are being required to perform work in areas that they may not be familiar with. Consider the purchasing agent who now actually needs to be involved with the initial selection of plastic material and press operators who must perform quality control activities as part of their machine operation responsibilities.

Companies are more aware that it is beneficial to train employees as much as possible and encourage them to take advantage of in-house as well as off-site training and seminars. In the future, these activities will become more prevalent as employees will be expected to cross-train and take on more and more responsibilities even as a condition of employ-ment. More emphasis will be placed on specialized education, and plas-tics courses will become more available at colleges and universities, to the point that one may earn a bachelor's degree, and even a master's degree, in plastics engineering.

Alliances

An *alliance* might be defined as combining talents, resources, or expertise in an attempt to make the allying parties more efficient and productive in their respective efforts. For instance, a large computer company may find it beneficial to form an alliance with a small (or medium-sized) circuit board manufacturer. The alliance gives the computer company access to circuit board manufacturing technology, and gives the circuit board com-pany the security and financial resources of the computer company. An alliance creates benefits for both parties.

In the future, alliances will become extremely popular as large compa-nies move to reduce their own assets and personnel levels, and smaller companies seek the resources available from large companies with mini-mum expenditure on their own part. The result will be an increase in the number of medium-sized companies resulting from expansion of smaller companies and a decrease in the number of large companies as a result of downsizing. This will produce increased sharing of technologies and information between allied companies and between groups of specific industries.

Quality

The forming of alliances, coupled with the need to become globally accepted, will drive companies of all sizes to improve their quality systems. The demand for products on a worldwide basis will require that companies submit to, and accept, global quality control standards and procedures. The 1990s saw the beginnings of such programs with the advent of ISO 9000 activities and the concept of *World Class*. While these are separate programs, they are similar in the results they achieve, notably that any company that meets the requirements of one will basically satisfy the requirements of the other. By meeting these requirements, a company will be able to produce a product or provide a service for any other company, regardless of where it is located geographically, without the tedious, time-consuming qualification procedures that are now expected. And, if a company is certified through ISO 9000 (or another global quality strategy program), it will automatically be certified to do business with any other company registered under the standard.

SUMMARY

John Wesley Hyatt launched the injection-molding industry in 1868 by winning a contest to find a substitute for ivory in making billiard balls.
- The first injection-molding machine was formally patented in 1872 by John and Isaiah Hyatt.
- With the advent of World War II, the plastic injection-molding industry burgeoned because of the need for inexpensive, mass-produced products.
- In 1946, John Hendry received a patent for his screw-style injection-molding machine. This revolutionized the industry and allowed faster cycles, evenly distributed heat, and lower energy costs, while providing a material melt that was homogeneous.
- The future of the injection-molding industry will bring improved materials, energy- efficient processes, increased use of desktop manufacturing, larger *and* smaller tooling, much shorter lead times and development cycles, greater use of computer programs, increased acceptance of recycling practices, greater emphasis on education and training, an increase in the concept of forming business alliances, and improved quality under World Class and ISO 9000 certification programs.

QUESTIONS

1. What prompted the birth of the plastic injection-molding industry?
2. Who was responsible for inventing the first injection-molding machine, and when?
3. Who invented the first screw-style injection-molding machine, and when?
4. What are some of the advantages of using a screw injection machine over a plunger machine?
5. What is the one major advantage to using a plunger-style machine?
6. What is the definition of *desktop manufacturing*?
7. How many material choices were available in 1995?

    (A) 180   (C) 18,000

    (B) 1,800   (D) 180,000

8. As a business concept, what is the definition of the term *alliance*?

The Molding Machine 2

THE MAIN COMPONENTS

Injection molding, the most popular process for manufacturing thermoplastic products, consists of injecting molten plastic material from a reservoir (heated cylinder) into a closed mold, allowing the plastic to cool down and solidify, and ejecting the finished product from the mold. The machine consists of an injection unit to inject the material, and a clamping unit that is used to hold the mold closed during the injection phase. This basic machine concept is shown in Figure 2-1.

The injection unit and the clamp unit serve separate primary purposes, and they complement each other. However, a machine may be purchased and built with virtually any combination of injection unit and clamp unit because each is independent of the other. There are some basic guidelines that help determine which injection/clamp combination is correct for specific applications. Table II-1 shows common clamp and shot size combinations.

The Injection Unit

Sizing the Injection Unit

Primarily, and ideally, the injection unit should be sized so that it contains two full cycles' worth of material. In other words, 50 percent of the capacity of the injection cylinder should be emptied each time a cycle is completed. This emptied capacity is referred to as the amount of *shot* a machine takes for each cycle because the material shoots into the mold during the injection phase. The 50-percent rule is ideal, but shot size should never be less than 20 percent or more than 80 percent of cylinder capacity.

For example, if the total amount of material that is used for one complete cycle is 2 oz (56.7 g), the ideal injection unit for that application would exist on a machine with a 4-oz (113.4-g) cylinder, because 50 percent of 4 oz equals 2 oz, which is the initial requirement. And, using the 20- to 80-percent minimum/maximum limits, the 2-oz shot could be produced on as small a machine as one with a 2 1/2-oz (71-g) cylinder (80 percent), and

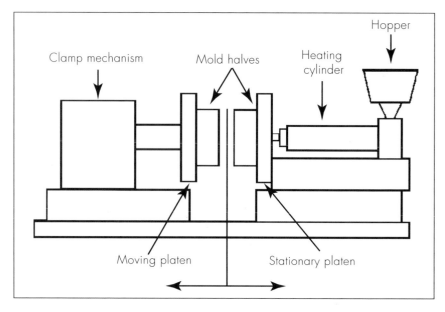

Figure 2-1. Injection-molding machine.

Table II-1. Determining Clamp and Shot Size Combinations

Clamp size, tons (kilonewtons)	Shot size, oz. (g)
10 (89)	1/2 (14.2)
25 (222.5)	2 (56.7)
50 (445)	4 (113.4)
100 (890)	8 (226.8)
200 (1780)	16 (453.6)
250 (2225)	20 (567)
300 (2670)	30 (851)
450 (4005)	60 (1701)
750 (6675)	120 (3402)
1000 (8900)	200 (5670)
2000 (17,800)	450 (12,757)
4000 (35,600)	900 (25,515)

on as large a machine as one with a 10-oz (283.5-g) cylinder (20 percent). What determines the shot size is the heat sensitivity of the specific material being molded. Some materials are very heat sensitive and burn easily, while others are much less heat sensitive and are able to withstand longer exposures to elevated temperatures. Heat sensitivity will be discussed later in this chapter and in more detail in Chapter 6. Heat sensitivity determines the amount of time the material is able to stay within the heated injection cylinder before it begins to degrade. Degraded material will not produce quality products. The 50-percent rule of thumb noted earlier ensures that no material, regardless of its allowed residence time, will degrade while being molded. The 20-percent rule of thumb applies to materials with low heat sensitivity, and the 80-percent rule applies to materials that are extremely heat sensitive.

Note that the capacity of an injection unit is rated in terms of the weight of polystyrene it can hold. A conversion is required to determine how much of any other plastic it can hold, and this is done by comparing specific gravity values. Specific gravity values are available from the material supplier and from many plastic encyclopedia sources. For instance, polystyrene has a published specific gravity value of 1.04. In the case of polycarbonate, the specific gravity value is 1.20. Specific gravity is an indicator of weight, with the higher values indicating heavier materials. The specific gravity (sg) value of the selected material (in this case, polycarbonate) is divided by the specific gravity value of polystyrene to determine how much of the selected material can be held in the cylinder. For this example, the polycarbonate sg of 1.20 is divided by the polystyrene sg of 1.04, giving a value of 1.15. This means that a machine capable of injecting 10 oz (283.5 g) of polystyrene could also inject 11.5 oz (326 g) of polycarbonate.

Purpose of the Injection Unit

The injection unit must perform many duties and contains many components that contribute to the performance of these duties. Figure 2-2 shows most of these components.

The Heating Cylinder

The heart of the injection unit is the heating cylinder, also called the *barrel*. It is usually manufactured in the form of a long, round tube and is made of an inexpensive grade of steel. The inside of the tube is lined, usually with a thin sleeve of high-quality hard tool steel that can withstand the abrasive nature of the injection process. Normally, the sleeve has a high chromium content.

The outside of the barrel has heater bands strapped to it. The bands are electrically activated and are placed along the entire length of the barrel with minimal space between them. Note in Figure 2-2 that there are three heater zones: rear, center, and front. Each zone contains three or more heater bands (depending on the length of the injection cylinder) and each zone is individually controlled by an electrical unit located in the control panel of

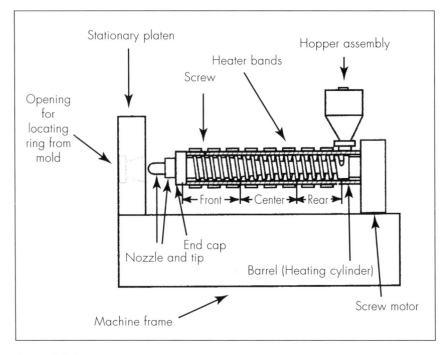

Figure 2-2. Injection unit components.

the machine. Each temperature control unit is fed temperature information by a thermocouple in a hole in the wall of the heating barrel in the area of the zone it is controlling. The control unit then decides whether more heat is required and, if so, energizes the heater bands in that zone. When the selected temperature is reached, the thermocouple informs the control unit, which stops sending electricity to the heater bands until the temperature drops again, at which point the cycle repeats.

Minimum and maximum temperature limits are set on the control unit and used by the unit to determine whether the heater bands should be energized or de-energized. A single control unit and thermocouple are assigned to a single heating zone, but there are three or more heaters in each zone, so each control unit actually controls three or more heaters at the same time. Because all of the heaters in a single zone are wired together, whatever the control unit does to one heater it does to all of the heaters in that zone.

The Basic Hopper

In the upper right-hand section of Figure 2-2 is a component called the *hopper*. This is where raw plastic pellets are stored before they are introduced to the heating cylinder. In Figure 2-3, it can be seen that this unit has tapered sides to facilitate dropping fresh material (by gravity) into the barrel. The hopper is designed to hold approximately 2 hours' worth of raw material for the specific machine. The amount is based on normal cycles and average part weights usually produced on a machine of that size.

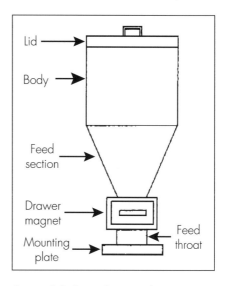

Figure 2-3. Basic hopper design.

The base of the hopper should contain a magnet—either an external drawer magnet (as shown) that can be pulled out and cleaned while the machine is running its normal cycles, or a loose magnet that is placed inside the hopper and must be pulled out of the hopper for cleaning. While the latter is less expensive, the former is much easier to clean. The purpose of the magnet is to trap any loose particles of metal that may have been placed in the raw plastic either by accident or intentionally. Accidental particles include small metal filings that come from the blades of a plastic granulator used to produce regrind or the metal mixing components used in the manufacture of the raw plastic pellets. Metal particles can ruin the sleeve of the injection barrel or the surface of the screw.

The Injection Screw

The screw is an auger-shaped rod that is placed inside the heating barrel. The primary function of the screw is to auger fresh material from the hopper area into the heating area of the barrel. A secondary function is to mix and homogenize the molten plastic. The screw also generates heating friction to raise the temperature of the plastic. The friction is created because there is just a slight clearance between the surface of the screw flights and the inside wall of the barrel, usually only 0.003 to 0.005 in. (0.008 to 0.013 cm). As the material is brought forward along the screw flights, the plastic is squeezed tighter and tighter. The friction of squeezing generates heat.

The external heater bands supply most of the heat for softening the plastic. The screw, however, does provide additional heat, and this reduces the amount of electricity required to heat the plastic completely. The squeezing action of the screw is called *shear*. Too much shear can tear up the plastic molecules and degrade the material, making it inferior or even useless. For this reason, the screw itself cannot be used to impart all of the heat needed.

Injection Screw Designs

There are many different screw designs, with various shapes of flights, distances between flights, amounts of shearing action, screw tip geometries, and methods of shutoff.

Figure 2-4 shows a typical screw design for an injection-molding machine. It is called a *metering screw*. The rear section (the *feed zone*) has a smaller screw diameter than the front end (the *meter zone*). The middle area (the *melt zone*) is a transition area between the meter and feed zones. Therefore, there is a smaller gap through which the plastic must flow. This results in a shearing action that creates the frictional heat mentioned.

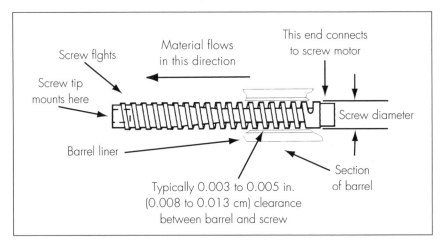

Figure 2-4. Typical injection screw design.

Screw Tip and Check Ring

Figure 2-5 shows how a typical screw tip mounts on the face of the injection screw. The tip itself is inserted through a check ring and seat designed to keep molten material from flowing back over the screw flights during injection. The tip fits into the face of the screw, usually with a left-hand thread to counteract the natural turning motion of the screw. A right-hand thread would tend to unscrew as a result of the turning action of the screw. The screw tip angle and length are determined by the viscosity of the plastic

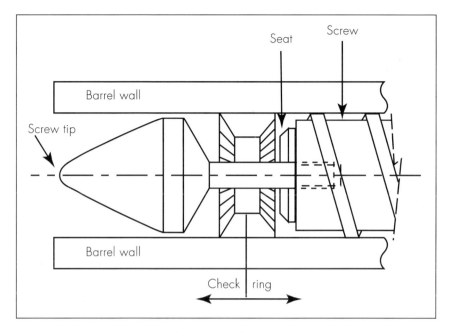

Figure 2-5. Screw tip and check ring assembly.

being molded. The material supplier or screw manufacturer can help make the final decision. In some cases, a general-purpose screw and tip can be utilized for a variety of similar materials, but it is better to use a specific design for a specific group of materials.

The usual injection machine uses a reciprocating screw. This simply means that the screw pushes forward and pulls backward (reciprocates), acting as a plunger to inject the molten plastic.

Nonreturn Valves and Ball Shutoffs

The purpose of the check-ring nonreturn valve mechanism in Figure 2-5 is to keep molten plastic from escaping back over the screw as the screw moves forward (acting as a plunger) to inject material into the mold. The check ring then is allowed to move forward as the screw augers fresh material forward to prepare for the next cycle. The action of the check ring allows that material to move in front of the screw tip. The sequence of the nonreturn is:

1. The screw pushes forward, injecting a charge of molten material into a mold.
2. The check ring is forced back against the screw tip seat and seals against it, preventing material from passing back over the screw.

3. The screw stops pushing and begins to turn (bringing new material forward).
4. The check ring slips forward under the influence of the pressure buildup.
5. Molten plastic flows into the space in front of the screw tip.

There are many different nonreturn mechanisms. The check-ring style is most common, but the ball-type device shown in Figure 2-6 is also popular.

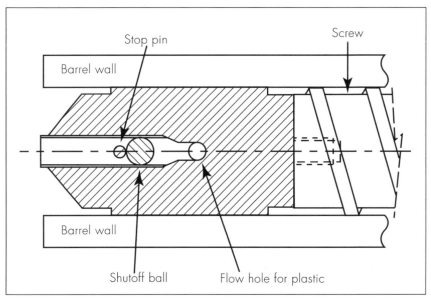

Figure 2-6. Ball-type nonreturn valve.

In this design scheme, the shutoff ball travels back and forth between the stop pin and the flow hole. When the screw moves forward (acting as a plunger), the ball moves backward, plugging the through hole and keeping material from flowing back over the screw flights. When the screw stops injecting and turns to auger fresh material forward, the ball is pushed forward, allowing material to flow and fill the space in front of the screw tip and into the machine nozzle (not shown).

In both the check-ring and ball-shutoff cases, the plastic material is restricted, even when the nonreturn devices are in the open flow position. With high-viscosity or heat-sensitive materials, this restriction may cause degradation of the plastic. So, nonreturn mechanisms are usually not used when molding these materials; in fact, they are not usually required due to the high viscosity of heat-sensitive plastics.

The Nozzle

One final item makes up the complete injection unit. The nozzle of the machine is a two-piece, tube-shaped component that bolts to the face of the injection barrel, as shown in Figures 2-7 and 2-8.

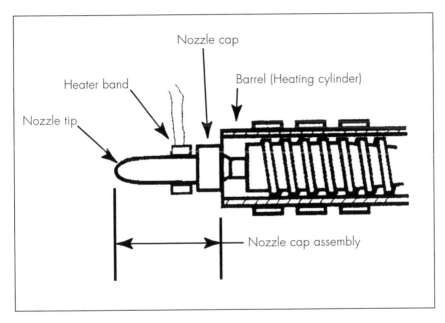

Figure 2-7. Typical nozzle assembly.

Note that the nozzle cap has an internal taper that matches that of the screw tip. Also, there is a tapered hole through the nozzle tip itself. The radius on the face of the nozzle tip fits up against a matching radius in the sprue bushing of the injection mold. Also, notice that there is a heater band on the nozzle tip. This is called the *nozzle heater* and it is controlled much like the other heater bands on the injection barrel.

There are some nozzle designs that incorporate shutoff devices in the form of needles, springs, sliding balls, or combinations of these. Their purpose is much the same as the nonreturn valve in the screw tip; they shut off the flow of plastic for those materials that are not highly viscous, such as nylon, and that tend to drool from standard nozzles.

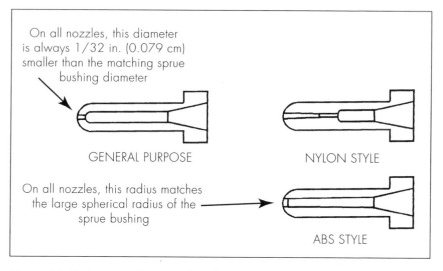

Figure 2-8. Various nozzle assembly styles.

Determining Injection Requirements

Calculation

A variety of injection requirements for a specific application can be determined mathematically. Some of these are addressed below.

Shear rate. Shear rate is defined as the surface velocity of the plastic at the wall of the heating barrel, divided by the depth of the screw flight channel, and the units are feet (meters) per minute. The formula is:

$$SR = (D \times N)/h,$$

where SR = shear rate
D = screw diameter
N = rate of screw rotation
h = depth of channel

An average shear rate value would be approximately 150 ft/min (45 m/min), but each plastic has a specific shear rate beyond which it will degrade. Heat-sensitive plastics such as polyvinyl chloride (PVC) have a lower shear rate (approximately 100 ft/min [30 m/min]) while nonsensitive materials may have a shear rate of 175 ft/min (53 m/min) or higher.

Shear rate values have a direct effect on the allowable speed of screw rotation. For example, from the above formula, it can be determined that using a standard 2-in. (5.1-cm) diameter screw with a material having an average shear rate of 150 ft/min would result in a maximum rotational speed of 230 rpm. Any rotational speed above that value will result in overshearing the plastic and thereby degrading it.

Screw output. The amount of material an injection machine can process is rated in pounds per hour (kilograms per hour). It is determined by how much horsepower is available to turn the screw. A 2-in. (5.1-cm)-diameter screw will normally withstand a maximum of 15 hp (11 kW). More than that may result in screw breakage. A 4.5-in. (11.4-cm)-diameter screw, on the other hand, will not break with up to 150 hp (110 kW) available to it.* For output, molding materials range from 5 to 15 lb/h for each horsepower (2.3 to 6.8 kg/h for each kW) applied. Therefore, a 15-hp system (2-in.-diameter screw) is capable of producing from 75 to 225 lb/h (34 to 102 kg/h) of plastic, depending on the viscosity (which affects shear rate).

Injection pressure. The average screw injection machine is capable of producing 20,000 psi (137,890 kPa) injection pressure in the heating barrel. This full pressure is available at the nozzle of the machine just before the material enters the mold. In most cases, it is advisable to use the highest injection pressure and the fastest injection speed possible to minimize the overall cycle time of the molding process. While 20,000 psi may be available, it is prudent to use only the highest amount of pressure required for a specific material and specific application. Normal practice is to begin molding at 6000 to 8000 psi (41,360 to 55,150 kPa) and increase/decrease pressure as necessary while optimizing the cycle. This is discussed in more detail later in this chapter and in Chapter 4.

L/D ratio. A critical factor involved in creating available injection pressure is the ratio of the length of the injection screw to its diameter (L/D). In Figure 2-9, note that the L dimension runs the entire length of the screw flights, and the D dimension goes over the largest diameter of the screw, which is also the overall flight diameter. The L dimension is normally 20 times greater than the D dimension. So, if the screw has a 2.5-in. (6.4-cm) diameter, the length of the flighted section should be at least 50 in. (127 cm). A 24:1 ratio is even better. In that case, a 2.5-in.-diameter screw would have a length of 60 in. (152.4 cm). The greater the ratio, the more gentle the shearing action of the screw on the plastic material.

The Clamp Unit

Sizing the Clamp Unit

The clamp unit of an injection-molding machine is rated by the maximum amount of clamp force the machine is capable of producing. This

---

*SPI Plastic Engineering Handbook, fourth edition (New York: Van Nostrand Reinhold, 1976).

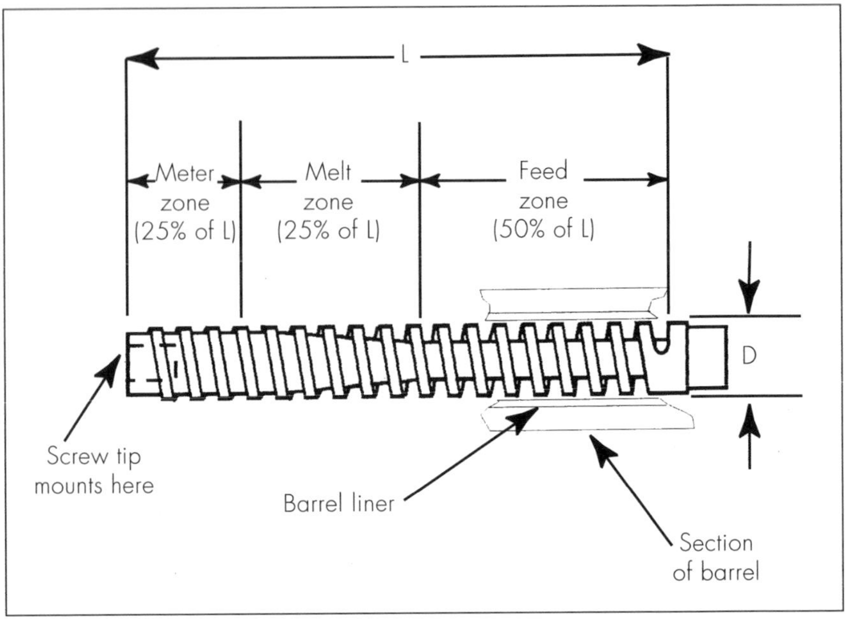

Figure 2-9. L/D ratio of screw.

force is needed to keep the mold closed during the injection process, which is the primary purpose of the clamp unit. Normally, the force rating is stated in tons (kilonewtons). So, a specific machine having a rating of 200 tons (1780 kN) is capable of producing a maximum clamp force equivalent to a total of 200 tons. But how much clamp force is necessary? And what happens if there is not enough clamp force? Or too much?

How Much Force Is Required?

The answer to this question depends on how much injection pressure is required to inject a specific plastic material into a mold. Chapter 6 addresses viscosity (the thickness value) of materials. This is a value that must be thoroughly understood; an explanation follows.

Thicker materials require greater injection pressures and are difficult to flow. Flow ranges in which each material will fall can be classified as *high flow, average flow,* and *low flow.* The melt flow index test determines the flow rate of any plastic, and the material suppliers make this information readily available on their material information sheets. These index numbers may range, for instance, from 5 to 20. The lower numbers signify that the specific material does not flow easily and would be classified as *low flow.* The higher numbers signify a material that flows very easily and would be classified as *high flow.*

It is not as important to remember a specific flow number as it is to know in what range a material falls: high flow, average flow, or low flow. Then, since more injection pressure is needed to inject a low-flow material than a high-flow material, it is understood that a low-flow material will require much more clamp force to keep the mold closed against that higher injection pressure.

A comparison of two materials will serve to demonstrate. A product molded of polycarbonate (a low-flow plastic) may require an injection pressure of 15,000 psi (103,410 kPa), while that same product molded of acetal (a high-flow plastic) may require only 5000 psi (34,470 kPa). Therefore, the polycarbonate product will require a clamp force on the mold that is approximately three times that for the acetal product.

Determining Projected Area

To determine the required clamp force, take the projected area of the part to be molded and multiply that number by a factor of from 2 to 8. Projected area is calculated by multiplying length times width. Figures 2-10 and 2-11 give an example.

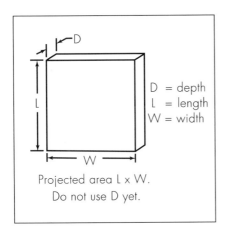

Projected area L x W.
Do not use D yet.

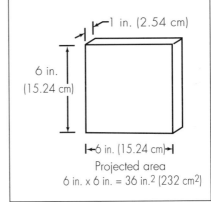

Figure 2-10. Determining projected area—A.

Figure 2-11. Determining projected area—B.

The projected area of the part is found by multiplying the *L* dimension by the *W* dimension (length × width). The *D* (depth) dimension is important only if it is more than 1 in. (2.54 cm). This is explained later. So, for this particular product, the projected area is determined by multiplying 6 in. × 6 in. (15.24 cm × 15.24 cm). The result is an area of 36 in.^{2} (232 cm^2). Clamp force requirements can now be calculated by multiplying the 36 square inches by a factor of between 2 and 8 tons per square inch (232 cm^2 × 27,580 to 110,320 kPa). The lower numbers can be used for high-flow materials and the higher numbers can be used for low-flow (stiff) materials.

For this example, polycarbonate has been selected as the material for molding. Polycarbonate is fairly stiff and a lower flow material, so the clamp factor used must be toward the high side. Experience has shown that a clamp factor of 5 tons/in.^{2} (68,950 kPa) is adequate for polycarbonate. That means that the 36 in.^{2} projected area found above must be multiplied by the clamp factor of 5 tons per square inch, to result in a total clamp tonnage requirement of 180 tons (36 × 5 = 180 [232 cm^2 × 68,950 kPa = 1600 kN]). There should be a safety factor of 10 percent added, so the final clamp force needed is 198 tons (1760 kN). The machine with the closest rating for this product would be a 200-ton (1780-kN) machine.

To summarize, the total clamp force required for a specific product is determined by finding the projected area of that product

$$\text{Projected area} = \text{length} \times \text{width}$$

and multiplying that area by a clamp factor of between 2 and 8 for the U.S. Customary System of Units (USCS). If in doubt, use 5:

$$\text{Projected area} \times 5 = \text{clamp force required} \quad (\text{USCS})$$

For the International System of Units (SI), the clamp factor ranges from 27,580 to 110,320. A general-purpose value is 68,950:

$$\text{Projected area} \times 68,950 = \text{clamp force required} \quad (\text{SI})$$

What About That D Dimension?

The D dimension becomes important only if the plastic part is more than 1 in. (2.5 cm) deep. That is not the thickness of the wall, but the total depth of the part. For every inch of depth over 1 in., the total clamp force must be increased by 10 percent. So, if the part shown in Figures 2-10 and 2-11 was 2 in. (5.1 cm) deep, the clamp force would be increased by 18 tons (160 kN) (10 percent increase for every inch over 1 in.) to a total of 198 tons (1760 kN). Add 10 percent for safety factor and the required force increases to 217.8 tons (1936 kN). The nearest machine size to that requirement would probably be a 225-ton (2002.5-kN) machine.

What Happens If Too Little Clamp Force Is Used?

If less than the calculated clamp force is used, the mold will not be able to stay closed when the plastic material is injected into it. The result will be flash, or nonfilled parts, or both.

Flash is material that squeezes out of a closed mold because injection pressure forces it out through any opening that allows material to flow. Sometimes flash is planned, but normally flash is unwanted because it creates an uncontrolled injection pressure situation and because it must be removed prior to shipping the finished product. Normally, flash removal requires a secondary operation, which adds cost to the product.

Nonfilled parts result when the mold opens up slightly (because of insufficient clamp pressure), keeping the prescribed amount of plastic from flowing into the entire shape of the mold.

What Happens If Too Much Clamp Force Is Used?

If too much clamp force is used, the injection mold, or the press, can be severely damaged. This damage can result from the collapse (or crushing) of the material used to make the mold (usually steel or aluminum). Molds can cost anywhere from a few thousand dollars to millions of dollars, and any damage is expensive to repair, if it can be repaired at all.

Another factor to consider is damage to the molding machine itself. Figure 2-12 depicts how a mold is mounted in a molding machine. The mold is actually mounted to two platens (pronounced *plattuns*), one stationary and the other moving. It is connected directly to the clamp mechanism. Mold half A stays with the stationary platen when the machine opens, but mold half B stays with the moving platen. Then the molded part can be ejected from the opened mold.

If there is too much clamp force, it is probably because the mold has been mounted in a machine that is too large. Therefore, the platens are much bigger than required, as shown in Figure 2-13. This results in the moving platen actually twisting and/or binding as it closes tight. The mold is not spread out enough across the face of the platen and this causes a lack of support to the platen. Constant opening and closing of the mold will result in the platen warping and/or bending while it moves. Eventually, the bushings and supports will wear out and the machine will lose its accuracy when closing and clamping the mold. So, by using too much clamp force, both the mold and the machine can be heavily damaged.

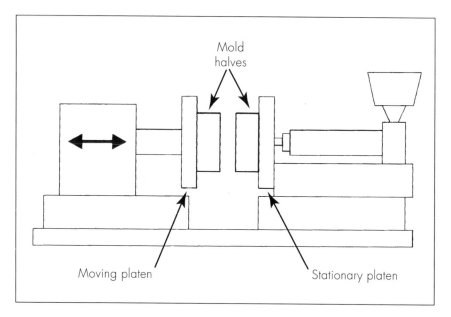

Figure 2-12. Typical mold mounted in press.

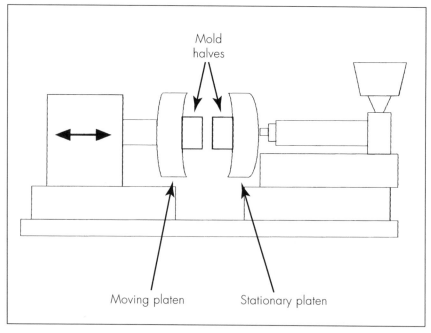

Figure 2-13. Small mold mounted in large press.

SUMMARY

The basic components of the injection-molding machine can be divided into two sections: the injection unit, rated in ounces (grams) of plastic available to inject, and the clamp unit, rated in total tons (kilonewtons) of clamping force available to keep the mold closed.

The injection unit consists of a hopper that feeds fresh material to the machine, a barrel that is heated by external heater bands, an augering screw that is mounted lengthwise in the heated barrel, and a nozzle that connects the injection unit to the mold.

The clamp unit consists of a clamping mechanism that is either mechanical or hydraulic, or both, and in some cases all-electric versions of mechanical. The clamp unit is used to keep the mold closed during the injection phase of the molding cycle. The amount of clamp force required is determined by the projected area of the product being molded.

QUESTIONS

1. What material is used as a standard for determining the capacity of an injection cylinder?
2. What percentage of this capacity should be injected during any single cycle?
3. How do you calculate the weight of one material versus another, knowing the specific gravity of both?
4. Name the three heater zones found in the injection barrel. Where is the fourth zone?
5. On the average, how often will a standard hopper require refilling?
   (a) 1 hour   (b) 2 hours
   (c) 4 hours   (d) 8 hours
6. How much pressure can the average molding machine generate in the injection cylinder?
7. What is the primary purpose of the clamp unit?
8. What is the formula for determining how much clamp force is required?
9. How is *projected area* determined?
10. What happens if:
    (a) excessive clamp force is used?
    (b) not enough clamp force is used?

Parameters of the Molding Process

3

IDENTIFYING THE PARAMETERS

Numerous variables affect the injection-molding process. In fact, a recent study itemized more than 200 different parameters that had a direct or indirect effect on the process.

Many years ago, I was asked by my manager to compile a list of all the various parameters associated with controlling the injection-molding process. At the time, I thought this would be an easy task and eagerly pursued the answers. I envisioned a short list comprising such things as injection pressure, cycle time, barrel temperature, and a few other common items.

I soon realized that I had grossly underestimated the number of parameters. For every parameter I found, another would appear. For instance, *injection pressure* consisted of more than one item. There were initial injection pressure, second- and up to fifth-stage injection pressure, holding pressure, back pressure, and line pressure, all of which had a direct effect on each other. Even items such as humidity and ambient temperature had an effect on the molding process. Shift changes, relief operators, fans blowing, housekeeping, age of equipment, size of machine, location of press, pressure of cooling water, all seemed to have direct or indirect effects on the injection-molding process.

I found, for example, that between the hours of 6 a.m. and 8 a.m., scrap rates increased dramatically and all the molding machines seemed to go out of control. Further investigation showed that, because we were located in a very small town, water pressure would drop at that time of the morning because of so many people getting ready for work. When the water pressure dropped, the cooling devices on our equipment did not function effectively and the machines would heat up. This had an effect on the overall cycle times, temperature of the operating oil, and temperatures of the individual molds. By the time adjustments were made to compensate for the unknown cause of the problem, the water pressure would return to normal and the changes that had been made had to be undone in order to return to normal operation.

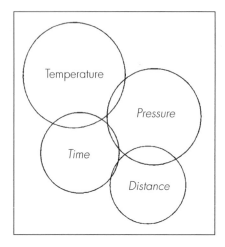

Figure 3-1. Main processing parameters.

Although there are so many different variables, it is not impossible to get control of the injection-molding process. What is needed is a more practical approach to understanding all of these parameters, and targeting those that have the greatest effect on the overall quality and cost-effectiveness of the finished molded product.

Figure 3-1 shows that all of the parameters involved can be placed into four basic categories: temperature, pressure, time, and distance. The relative importance of the categories is shown by the size of the circles. Thus, temperature is the most important, followed by pressure, time, and distance. However, each is dependent on the other, and changing one will affect one or all of the others. The discussion that follows addresses that interdependence.

TEMPERATURE

A variety of temperatures affect the injection-molding process, ranging from melt temperature to mold temperature, and including even ambient temperature.

Melt Temperature Control

Melt temperature is the temperature at which the plastic material is maintained throughout the flow path. This path begins where the plastic material is transferred from the machine hopper into the heating cylinder of the injection unit. Then the material is augered through the heating cylinder and into the machine nozzle. From there it is injected into the mold, where it must travel along a runner system (if one exists), through the gates, and into the cavities that are machined into the mold. The temperature of the melt must be controlled all along the path, starting with the heating cylinder.

Figure 3-2 shows that the heating cylinder is wrapped with heater bands. These are electrical heaters shaped like hinged bracelets that mount around the outside of the heating cylinder. There are three main heating zones to the heating cylinder: the *rear zone*, the *center zone*, and the *front zone*. In

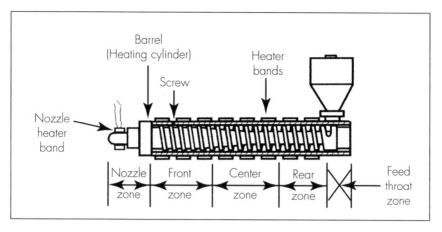

Figure 3-2. Heating cylinder (barrel).

addition, there is usually at least one heater band fastened around the machine nozzle, an area referred to as the *nozzle zone.*

The plastic for the injection process should be brought up to proper temperature gradually as the material drops from the hopper into the rear zone, where the initial heat begins to soften the material. Then the material is augered forward by the screw into the center zone where the temperature is generally 10 to 20° F (5.6 to 11° C) higher than in the rear zone. As the material travels to the front zone, the temperature is again increased by 10 to 20° F, and finally the material is ready to be injected into the mold. It is held at this point until the previous molding cycle is complete, at which time the mold opens, parts are ejected, the mold closes, and the next cycle begins. The charge of preheated plastic is then injected into the mold.

Besides absorbing heat from the externally mounted heater bands, the plastic material absorbs a large amount of heat from the friction caused by the augering action of the injection screw. (More than 95 percent of all injection machines have reciprocating screws as opposed to the standard straight plunger design of years gone by.) The screw rotates to bring fresh material into the heating cylinder and prepare it for the coming cycle. While being pulled along, the new material is squeezed between the flights of the screw and the inside wall of the injection barrel. The friction generates heat, which is absorbed by the plastic.

The main point here is that the plastic must be heated to the proper temperature for injection. Table III-1 lists the melt temperatures of some common plastics. Melt temperature is measured at the nozzle as the plastic exits the machine, before it enters the mold. It is measured by taking an

Table III-1. Suggested Melt Temperatures for Various Plastics

Material	Temperature, °F (°C)
Acetal (copolymer)	400 (204)
Acetal (homopolymer)	425 (218)
Acrylic	425 (218)
Acrylic (modified)	500 (260)
ABS (medium-impact)	400 (204)
ABS (high-impact and/or flame retardant)	420 (216)
Cellulose acetate	385 (196)
Cellulose acetate butyrate	350 (177)
Cellulose acetate propionate	350 (177)
Ethylene vinyl acetate	350 (177)
Liquid crystal polymer	500 (260)
Nylon (Type 6)	500 (260)
Nylon (Type 6/6)	525 (274)
Polyallomer	485 (252)
Polyamide-imide	650 (343)
Polyarylate	700 (371)
Polybutylene	475 (246)
Polycarbonate	550 (288)
Polyetheretherketone (PEEK)	720 (382)
Polyetherimide	700 (371)
Polyethylene (low-density)	325 (163)
Polyethylene (high-density)	400 (204)
Polymethylpentene	275 (135)
Polyphenylene oxide	385 (196)
Polyphenylene sulfide	575 (302)
Polypropylene	350 (177)
Polystyrene (general purpose)	350 (177)
Polystyrene (medium-impact)	380 (193)
Polystyrene (high-impact)	390 (199)
Polysulfone	700 (371)
PVC (rigid)	350 (177)
PVC (flexible)	325 (163)
Styrene acrylonitrile (SAN)	400 (204)
Styrene butadiene	360 (182)
Tetrafluoroethylene	600 (316)
Thermoplastic polyester (PBT)	425 (218)
Thermoplastic polyester (PET)	450 (232)
Urethane elastomer	425 (218)

"air shot" and plunging a probe from a measuring instrument with a fast response time (1 second is acceptable) into the plastic melt. An air shot is made with the injection sled pulled back so the injection unit does not touch the mold. The material is then released as in a normal cycle, but it is injected into air rather than the mold. It is allowed to fall onto a tray made for the purpose and its temperature is then quickly measured. The temperature at that point should be within 10° F (5.6° C) of the desired temperature.

Insulation Blankets

To better regulate and control the temperature of the injection barrel (cylinder), an insulation blanket is used. This is a nonflammable jacket that fits around the outside of the heating cylinder, directly over all the heater bands, and keeps heat from being lost to the atmosphere. It makes little sense to have the heater bands supply a great amount of heat to warm up plastic inside the barrel while much of that heat is lost to the surrounding air. With insulation blankets, heat generated by the heater bands is directed only toward the barrel; less energy is required to heat the plastic, and operating costs are lower. Use of blankets can reduce energy costs to heat the plastic by 25 percent or more.

Mold Temperature Control

The plastic material is now ready to flow into the mold. First, it must travel through the machine nozzle, which is the last heating zone provided by the machine. After the material exits the nozzle and enters the mold, it immediately begins to cool down as the mold absorbs heat from it. The rate at which this heat is absorbed determines how far the plastic will flow before it begins to solidify and stop moving. Each product, depending on its design and plastic material, demands specific cooling rates, and this rate of cooling is critical to product quality. Therefore, the mold temperature must be regulated in order to regulate the cooling rate of the plastic. This is done by connecting the mold to a temperature control unit that normally utilizes water as a medium. The water is circulated through the mold and held at a preset temperature by heating or cooling in cycles.

Every combination of plastic and product has a specific temperature at which the mold should be maintained to ensure quality molding, but Table III-2 suggests starting points for a variety of common plastics.

The mold temperature is measured directly from the molding surface of the tool with a solid probe on a pyrometer device. Usually, readings from several areas are averaged. Remember that these temperatures are recommended starting points only and should be adjusted for specific applications.

Table III-2. Suggested Mold Temperatures for Various Plastics

Material	Temperature, °F (°C)
Acetal (copolymer)	200 (93)
Acetal (homopolymer)	210 (99)
Acrylic	180 (82)
Acrylic (modified)	200 (93)
ABS (medium-impact)	180 (82)
ABS (high-impact and/or flame retardant)	185 (85)
Cellulose acetate	150 (66)
Cellulose acetate butyrate	120 (49)
Cellulose acetate propionate	120 (49)
Ethylene vinyl acetate	120 (49)
Liquid crystal polymer	250 (121)
Nylon (Type 6)	200 (93)
Nylon (Type 6/6)	175 (79)
Polyallomer	200 (93)
Polyamide-imide	400 (204)
Polyarylate	275 (135)
Polybutylene	200 (93)
Polycarbonate	220 (104)
Polyetheretherketone (PEEK)	380 (193)
Polyetherimide	225 (107)
Polyethylene (low-density)	80 (27)
Polyethylene (high-density)	110 (43)
Polymethylpentene	100 (38)
Polyphenylene oxide	140 (60)
Polyphenylene sulfide	250 (121)
Polypropylene	120 (49)
Polystyrene (general purpose)	140 (60)
Polystyrene (medium-impact)	160 (71)
Polystyrene (high-impact)	180 (82)
Polysulfone	250 (121)
PVC (rigid)	140 (60)
PVC (flexible)	80 (27)
Styrene acrylonitrile (SAN)	100 (38)
Styrene butadiene	100 (38)
Tetrafluoroethylene	180 (82)
Thermoplastic polyester (PBT)	180 (82)
Thermoplastic polyester (PET)	210 (99)
Urethane elastomer	120 (49)

The object of the cooling process is to lower the temperature of the molded plastic to the point at which it solidifies again. When the plastic reaches that point, it can be ejected from the mold with relative structural safety. That simply means that the plastic part will not move excessively, causing warpage, twisting, or other shrinkage-related problems as the plastic continues to cool.

Postmold Shrinkage Control

Although molded thermoplastic products appear to be stable, they will continue to cool and shrink for up to 30 days after being ejected from the mold. Most (95 percent) of the total shrinkage will occur during the time the plastic is cooling in the mold. The remaining 5 percent will take place over the next 30 days, but most of that will happen within the first few hours after ejection from the mold. Thus, it is important to inspect molded parts after they have been allowed to stabilize. Initial inspection can be performed as soon as the part cools to the touch after being ejected. But more accurate inspection can only be performed after the part has cooled for 2 to 3 hours or more.

Hydraulic System Temperature Control

Besides melt temperature and mold temperature, there is the temperature of the hydraulic system of the press to be considered. The temperature of the hydraulic oil in these systems must be maintained between 80 and 140° F (27 and 60° C), in most cases. If the oil is too cool, it will be thick (viscous) and cause sluggish action of hydraulic components. If it is too hot, it will break down, causing components to stick or valves to malfunction. The temperature of the oil is regulated by a heat exchanger mounted on the injection machine. This heat exchanger acts like a radiator on a car and cools the oil by circulating it around tubes filled with circulating water. These tubes must be kept clean and require periodic flushing with an acid cleaner. If the oil is allowed to overheat, that heat will eventually transfer throughout the entire machine, including the platens to which the mold is mounted. This will cause the mold to overheat and result in poor-quality parts.

Ambient Temperature Control

Ambient temperature is also a concern. A particular job may be running perfectly well until someone opens a loading dock door or turns on a cooling fan in the vicinity of the molding press. This causes a change in the temperature of the air surrounding the machine and this, in turn, results in fluctuations in the readings provided by the various temperature control units of the machine. The injection process then becomes unstable for a period of approximately 2 hours, assuming no other changes occur to alter the ambient conditions. If more changes do occur, the process is unstable for longer periods of time.

Insulation Sheets

One method of reducing energy costs and controlling temperatures is to install *insulation sheets* on the outer surfaces of the injection mold (Figure 3-3). These sheets are available from a variety of sources, such as injection-molding accessories dealers and mold component suppliers. At one time they were made of asbestos, but since the ban on that material they are usually made of thermosetting polyester. They are available in different thicknesses, the proper one being determined by the size of the mold being insulated. Common thicknesses are 1/4 and 3/8 in. (0.64 and 0.95 cm).

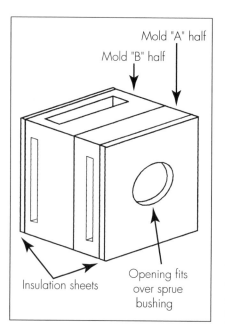

The sheets are cut to fit over all exposed outer surfaces of the mold. All six sides should be covered. The sheets are then drilled and countersunk to accept flathead screws, which will keep the sheet surface unobstructed. The sheets are cut to fit around waterline connections, locating rings, and other items typically mounted on the mold's outer surfaces. Then they are mounted directly to the mold surfaces with flathead screws. The sheets will compress very slightly under clamping pressures, but will quickly take a final set.

Figure 3-3. Mold insulation sheets.

The greatest benefit of using insulation sheets is that they create a much smaller area needing to be temperature controlled. Without the sheets, the atmosphere absorbs much of the heat of the exposed mold, causing the temperature control unit to supply more heat (or cooling) to maintain the proper mold temperature. Also, there is greater fluctuation in the temperature of the mold because of changes in ambient temperature and airflow around the mold. This requires greater and more frequent exercise of mold temperature control, which incurs higher energy use and higher costs. The insulation sheets create an environment that includes only the mold itself; therefore, that is all that needs to be controlled by the temperature control units. A higher degree of control can be attained, which results in lower temperature excursions, resulting in greater efficiency and lower energy costs. In fact, in comparing energy costs between two identical molds, one with insulation sheets and one without, the mold with insulation sheets showed a 25 percent savings in energy costs. Insulation sheets also minimize (or eliminate) sweating of the mold caused by high-humidity conditions.

PRESSURE

There are two areas in the injection machine that require pressure and pressure control: the injection unit and the clamp unit. They are closely related in that they are opposing pressures—the clamp unit must develop enough clamp pressure to overcome the pressure developed by the injection unit during the molding process.

Injection Unit

Three basic types of pressure are developed by the injection unit: initial pressure, hold pressure, and back pressure.

Initial Injection Pressure

This is the first pressure that is applied to the molten plastic. It develops as the result of main system hydraulic pressure pushing against the back end of the injection screw (or plunger) (Figure 3-4).

The amount of pressure developed by the main system is on the order of 2000 psi (13,789 kPa). Some systems are capable of producing more than that, but 2000 psi is the most common line pressure. This pressure is converted to a maximum of 20,000 psi (137,890 kPa) at the nozzle of the injection unit (where the plastic first enters the mold) by the design and shape of the injection screw. In most cases, the full 20,000 psi is not required for filling a mold, and most products can be molded in a range of from 5000 to 15,000 psi (34,472 to 103,418 kPa). The pressure actually required depends on the plastic being molded, the viscosity and flow rate of the plastic, and the temperatures of the plastic and the mold.

The ideal situation is to be able to fill the mold initially with the highest practical pressure in the shortest practical time. Normally, the initial fill can be accomplished in less than 3 seconds. Note that even though "the

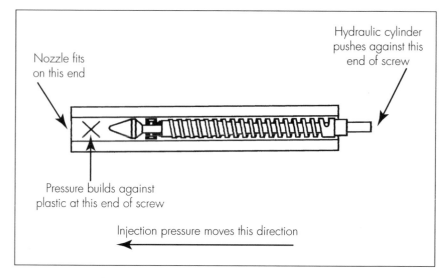

Figure 3-4. Developing injection pressure.

highest practical pressure" should be used, a constant effort should be made to keep that practical pressure requirement low so molded-in stresses are minimized.

To summarize, initial injection pressure is used to create the initial filling of the mold. It should be set at the highest practical value to fill the mold with the fastest practical speed.

Holding Pressure (Secondary Pressure)

This pressure is applied at the end of the initial injection stroke (Figure 3-5) and is intended to complete the final filling of the mold and hold pressure against the plastic that was injected so it can solidify while staying dense and "packed." As a rule, the amount of pressure used here can be half the initial injection pressure or less. So, if initial pressure was 12,000 psi (82,734 kPa), the holding pressure can be approximately 6000 psi (41,367 kPa). The holding pressure is actually applied against a *cushion* or *pad* of material, which is discussed under "Distance" later in this chapter.

To summarize, holding pressure is used to finish the filling of the mold and pack the plastic material into the cavity image.

Back Pressure

Back pressure is applied after the injection phases mentioned above. When the hold pressure phase is completed, a signal is sent to the machine to start turning the screw to bring new material to the front of the barrel in

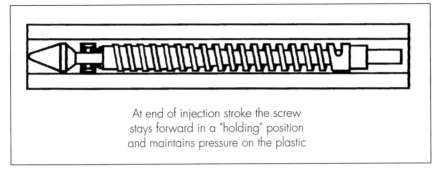

At end of injection stroke the screw
stays forward in a "holding" position
and maintains pressure on the plastic

Figure 3-5. Applying holding pressure.

preparation for the next cycle, or *shot* (so called because plastic shoots into the mold). The screw is not *pulled* back. Instead, the churning, or augering, action of the screw brings new material forward, and as that material fills up in front of the screw, the material itself begins to *push* the screw backward (Figure 3-6).

The back pressure is small compared to the injection pressure. A minimum of 50 psi (345 kPa) and a maximum of 500 psi (3447 kPa) is all that is required. The proper method of determining the amount of back pressure is to begin at 50 psi and increase, only if necessary, in increments of 10 psi (69 kPa) until the proper mix and density are achieved. Use of back pressure helps ensure consistency in part weight, density, and material appearance. It also helps to squeeze out any trapped air or moisture not eliminated by predrying the material. This minimizes (or even eliminates) voids in the molded product. If less than 50 psi back pressure is attempted, the controls and gages are not consistent nor accurate enough to maintain or indicate the actual pressure being developed. Thus, faulty readings and settings can occur. If more than 500 psi back pressure is attempted, the screw may not return at all, or it will stay forward much too long, and the plastic material will degrade under the extreme shear imparted to it from the continued churning action of the screw. In the case of reinforced plastics (such as glass-filled), the reinforcement material will break down, and this results in much less strength than is required in the molded product.

Clamp Unit

The purpose of developing clamp pressure is to keep the mold clamped shut against the forces developed when injection pressure pushes plastic into the closed mold. Therefore, the amount of clamp force must be at least equal to the amount of injection force.

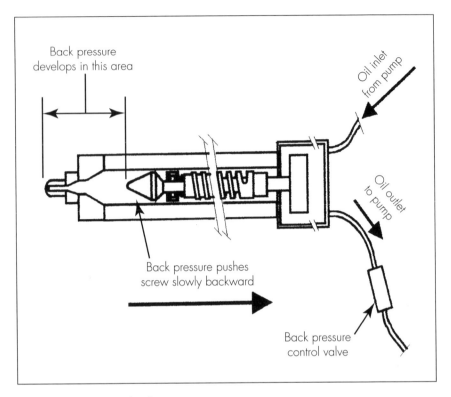

Figure 3-6. Applying back pressure.

Clamp pressure is applied to the mold either hydraulically or mechanically. There are advantages and disadvantages associated with each method.

Hydraulic Clamp System

In this method, the clamping force is developed by a hydraulic cylinder. A piston from the cylinder is attached to a moving platen on which the mold is mounted (Figures 3-7a and b).

The greatest advantage of this type of clamp system is that the clamp pressure can be regulated over a wide range. For instance, if the machine is rated at a 250-ton clamp force, the clamp force can be set anywhere from approximately 50 tons to the full 250 tons (445 to 2225 kN). This allows the proper clamp tonnage to be used for the specific job and minimizes the amount of energy expended. Using more tonnage than necessary not only wastes money, but may cause extensive damage to the mold or the machine or both because of the crushing forces applied.

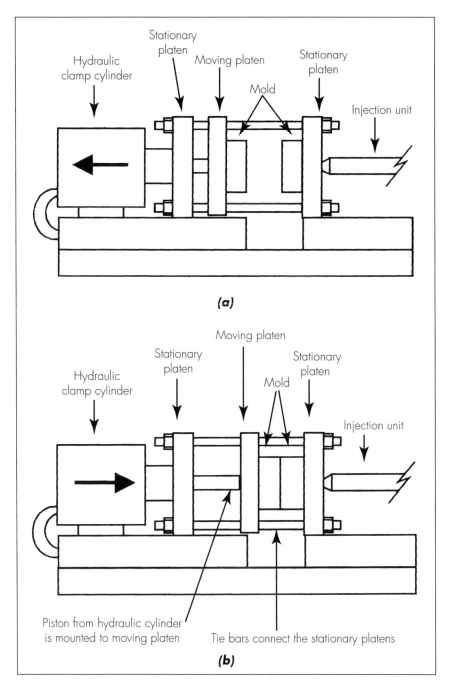

Figure 3-7. Hydraulic clamp; open (a), closed (b).

The greatest disadvantage of the hydraulic clamp is that when tonnage requirements approach the maximum rating, extreme injection pressures may overcome the clamp force and blow the mold open. For instance, if a mold requiring 225 tons is placed in a 250-ton machine and the injection pressure is on the high side (15,000 psi [103,410 kPa] or more), the potential exists for the injection pressure to overpower the clamp pressure, in which case the mold will open while plastic is being injected. This results in flash, short shots, and possible cycle interruption.

Mechanical Clamp (Toggle) System

The mechanical system utilizes a *knuckle and scissors (toggle)* mechanism to close the mold. The toggle is attached to the moving platen on which the mold is mounted. When the clamp is open (Figure 3-8*a*), a small hydraulic cylinder actuates the arms by pushing along their centerline. As the piston moves forward, it pulls the arms together, closing the mold (Figure 3-8*b*).

For the mold to close under full tonnage, the knuckles must actually pass center to lock. If they do not lock, they will not hold in the forward position and the injection pressure will blow the mold open. This can be demonstrated by watching a person push an arm straight out from the body. When the elbow is past center and straight, the arm is locked in the forward position and is difficult to push back until the elbow is relaxed.

The principal advantage of the mechanical system is that once it is locked in place, it is virtually impossible to blow the mold open even if injection pressures are beyond those required. Of course, there are limits to the pressure it can sustain, and eventually machine damage will occur if the injection pressures are held beyond requirements for extended periods. But once the system locks, there is no doubt that full tonnage force is available.

There are two distinct disadvantages to this system. First, there is considerable wear on the knuckle linkages and bushings must be replaced regularly. Second, there is little accommodation for adjustment on this system. If the machine is rated at 250 tons, the only tonnage available is 250 tons. It cannot be reduced, except minimally. Thus a smaller, borderline mold could not be run in this press without the distinct possibility of damage from crushing.

Some machines combine both hydraulic and mechanical systems for mold clamping, and some even incorporate electric motors for performing the mechanical action instead of hydraulic cylinders.

How Much Pressure is Needed?

As mentioned, total clamp force needed is determined by the projected area of the part being molded. This projected area is multiplied by a clamp

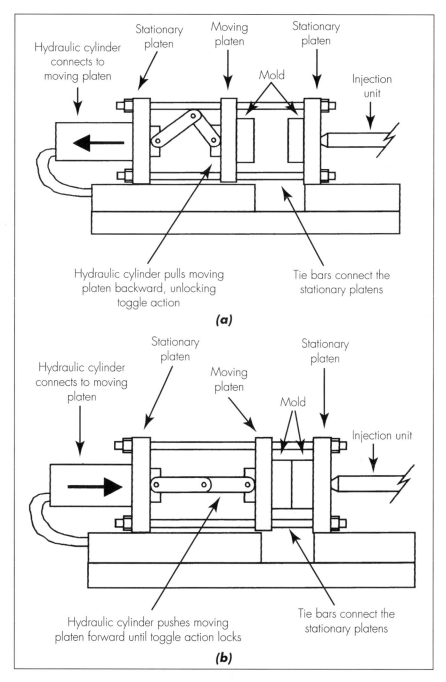

Figure 3-8. Mechanical clamp; open (a), closed (b).

force of from 2 to 8 tons for each square inch (27,580 to 110,320 kN/m²) of that projected area. As a rule of thumb, 4 or 5 tons/in.² (55,160 to 68,950 kN/m²) can be used for most products. If the plastic material is very stiff, it will require more injection pressure to fill the mold, thus more clamp tonnage to hold the mold closed. Conversely, if the plastic material has good flow characteristics, less injection pressure will be needed to fill the mold and a lower clamp tonnage will be adequate to keep the mold closed.

TIME

Gate-to-gate Cycle Time

During the injection-molding process, many internal activities are taking place. Some occur simultaneously with others (parallel activities), and some must wait until others are completed (serial activities). The overall cycle time provides a measure of the time required for all these activities. This is usually referred to as the *gate-to-gate* cycle time because it is common to start timing the overall cycle as soon as the machine operator closes the safety gate of the machine. The timing continues until the operator closes the same gate to start the next cycle. The entire amount of time elapsed between these two actions is the gate-to-gate or overall cycle time. (Timing of the overall cycle can actually be started at any point in the cycle as long as the timing continues until that same point in the next cycle.) The cycle time provides the only way to get an accurate picture of how long it takes to mold a product. This number is then used to determine the actual cost involved to manufacture the product. This, in turn, is used to determine the selling price of the product. Determining molding costs will be discussed later in this chapter.

According to the true gate-to-gate concept, the total overall cycle starts with the operator closing the safety gate and includes the activities listed in Table III-3 with typical time estimates. While the numbers add to 42 seconds, the actual total cycle is approximately 30 seconds because some operations are being performed during the time that others are operating, so there is an overlap. Each of these items is described and discussed in Table III-3.

Gate Close Time

Gate close time is the time it takes for the machine operator to actually close the safety gate, which starts the cycle. Each person operates a little differently, so the gate close time may fluctuate from operator to operator. Also, although a molding job may be set up and running fine, as soon as a

Table III-3. Average Times for Cycle Activities

Parameter	Average Value
Gate closing time	1 second
Mold closing time	4 seconds
Initial injection time	3 seconds
Injection hold time	5 seconds
Cooling time	12 seconds
Screw return time	8 seconds
Mold open time	4 seconds
Ejection time	1 second
Part removal time	2 seconds
Mold inspection, clean, spray, etc.	2 seconds

relief operator is placed on the job, the cycle begins to change. This does not mean the relief operator is not performing the job properly. It simply shows that each person operates at a slightly different pace, and this must be taken into consideration when setting up and running a job. Thus, the timing of the gate closing must be controlled as consistently as possible. The operators should be trained and informed that any slight change in the pace at which they close the gate may greatly affect the overall machine cycle. In fact, an increase of 2 seconds in the average cycle time of 30 seconds can cost approximately $20,000 annually, depending on the number of cavities, hourly wages, and cost of utilities. That additional cost must be paid for by the molder because the customer is not responsible for the increased cycle time. Of course, the opposite is also true: if the cycle can be reduced by 2 seconds, the molder can realize an additional profit of that same $20,000 annually. This shows the importance of maintaining consistent cycles and developing realistic cycles.

Mold Close Time

Mold close time is the amount of time it takes for the moving half of the mold to travel the entire distance to meet the stationary half of the mold, and lock up with full clamping force. This motion is usually initiated by the closing of a limit switch when the operator closes the safety gate to start the cycle.

There are actually two mold closing phases. The first is the initial closing that quickly brings the two mold halves together under low pressure. This takes approximately 1 to 2 seconds. But this action stops when the mold halves come to within half an inch (a centimeter) or less of fully closing. At that point, the speed slows down. This is a safety feature that keeps the mold from closing all the way before going into high pressure, should there be an obstruction within it, such as a broken-off plastic part from the previous cycle. It also allows any slides, cams, or other "actions" to operate slowly without danger of crushing. This final closing normally takes from 2 to 3 seconds.

Initial Injection Time

When the mold closes completely, either a limit switch or pressure buildup (or both) signals the injection screw to push forward and inject the molten plastic into the closed mold. The screw does not turn at this point, but only acts as a plunger to force the material into the mold. This initial injection is performed at the highest practical pressure for the specific application (normally 10,000 to 15,000 psi [68,940 to 103,410 kPa]) in the fastest practical amount of time. In most cases, the time is less than 2 seconds and rarely more than 3 seconds. Sometimes, depending on machine design, this action is divided into two or three smaller actions. Then, the total injection time normally does not exceed 4 to 5 seconds. The initial injection time is controlled by a timer. If booster injection is available, it will be included in the first stage of the initial injection time. When a booster phase is included, the injection machine's entire hydraulic system (injection and clamp) is combined to push a large volume of oil through the system. This can greatly increase the speed at which the material is injected into the mold.

Injection Hold Time

On most machines, the timer for initial injection time (also called *injection forward time*) controls the total amount of time that the injection screw is pushing forward. The initial injection time is the first part of that time, and injection hold is the latter part. On some machines, the hold time and initial time are on separate timers.

The hold time is the amount of time the injection screw maintains pressure against the plastic after it has been injected into the mold. This pressure is applied against the cushion or pad long enough for the gate to freeze off (solidify). The molten plastic enters the mold cavity image through a gate. The gate is the first point at which the plastic actually "sees" the cavity image. Once all the required material goes through the gate and packs the cavity image, the plastic is allowed to cool under hold pressure, down to the point at which it all solidifies. But, because it is normally the thinnest part of the cavity image, the gate is the first thing to solidify. When it does, there is no reason to maintain pressure because the plastic that is in the cavity lies beyond the solidified gate and the pressure from the injection unit no longer has any effect on it. So, pressure is held against the gate only long enough for the gate to freeze. In most cases, this is only a matter of a few seconds. A gate with a thickness of 1/16 in. (0.16 cm) would take approximately 6 seconds to solidify.

Cooling Time

Cooling time is probably the most important time in the entire injection process. It is the amount of time required for the plastic material to cool down to the point at which it has solidified *and* an extra amount of time to allow the plastic part to become rigid enough to withstand the ejection process (in which the finished molded product is pushed out of the mold after the entire cycle is completed). Even though the plastic may cool enough to solidify, it may not be rigid enough to be ejected. This is because the curing process actually takes as long as 30 days to finalize. The initial curing is rapid, and 95 percent of the total curing takes place in the mold. But the other 5 percent takes place outside the mold. If the outer skin of the plastic product is solidified to a sufficient depth, the remaining cooling will not have an appreciable effect on the molded part. But if the skin is too thin, the remaining cooling will cause shrinkage stress to build up and the molded part may warp, twist, blister, or crack.

The key to minimizing these problems is to keep the part in the mold for a sufficiently long time, but no longer than necessary because time is money, and long cycles are expensive. Most material suppliers are more than happy to share cooling time requirements for their materials at varying thicknesses (the thicker the part, the longer the cooling time required), but, on the average, a 1/16-in. (0.16 cm) thick wall should take approximately 9 to 12 seconds to solidify (depending on material) to the point at which it can be ejected from the mold without undue distortion.

Screw Return Time

After the screw pushes the plastic into the mold and holds it there until the gate freezes, the screw is ready to start turning to auger fresh material into the heating cylinder in preparation for the next shot. Although there is not usually a specific timer that can be used to adjust how long the screw should turn, the turning must take place before the cooling timer (sometimes called the *dwell timer*) times out. If the screw has not turned for a long enough time, there will not be enough material ready for the next shot. If the cooling timer times out, the mold may not be allowed to open until the screw finishes its travel.

The amount of time required for the screw to turn and return to the inject position is determined by how much back pressure is being applied (the more back pressure, the longer it takes the screw to return) and how much material needs to be prepared for the next shot. If the machine is the correct size for the job and the back pressure is properly set, the screw return time should be a few seconds less than the total cooling timer setting. In most cases, this amounts to approximately 6 to 8 seconds. The point at which the screw stops its return travel is determined by a limit switch. This switch is set to a point that ensures there is enough material ready for the next shot. As the screw turns (augers), it is pushed back toward its starting point by the pressure built up in front of it by the incoming material that augers forward. The screw continues to turn until it has traveled back and touches the preset limit switch, at which point the turning stops. Except for continuous purging, this is the only time that the screw actually turns (brings material forward). It does not turn during injection (unless something is wrong with the machine), except under rare circumstances when the turning is used to force more material into the mold. This is usually attempted only because the mold was not placed in a machine of correct size; it is a practice that should be avoided. If the screw does turn during injection, it usually means that the check ring or shutoff mechanism is worn or broken and should be replaced.

Mold-open Time

Mold-open time is the amount of time it takes for the mold to open. It is not determined by a timer (unless the machine is running fully automatically), but rather by the distance required for the mold to fully open and the speed at which it does it.

Commonly the mold opens in two stages. The first stage is very slow and the travel is short. This allows the vacuum created during the molding process to dissipate partially. When the plastic is injected into the mold, it displaces any air that is trapped in the closed mold. When air is displaced, a vacuum occurs. This vacuum causes the two mold halves to want to cling to each other. If the vacuum is not released, the two halves stay together when the machine opens, resulting in broken clamps and damaged molds and equipment. So, the vacuum is allowed to partially release by opening the mold slowly. After opening approximately 1/4 in. (0.6 cm) to relieve some of the vacuum, the mold is allowed to open fully. If there are fragile cams, slides, or lifting devices in the mold, it may be necessary to open slowly for a longer distance than 1/4 in. This will allow those devices to operate without slamming or shuddering. Then the mold is allowed to open fully. The speed of this final opening is much greater than the initial opening and can be set for as fast as the machine allows.

Ejection Time

When the cycle is completed and the mold has opened fully, the ejection system is allowed to come forward and knock the parts out of the mold. This action is normally started by a limit switch that actuates upon the full opening of the mold. However, sometimes it is performed through mechanical stops and actions. The ejection stroke itself is normally controlled by another limit switch that actuates when the right amount of ejection has taken place, but the speed at which the system comes forward must be controlled and this is what determines the amount of time required. There is still a partial vacuum in the cavity images. Therefore, the ejection system is pushed forward (usually by a small hydraulic cylinder) at a slow enough rate to overcome the vacuum, but fast enough to be practical. Ejection normally lasts for 1 or 2 seconds, depending on the length of ejection necessary.

The ejection system then must return before the next cycle can start. In some cases, it is not necessary to return the system because the closing of the mold will perform that action. Such practice is not recommended, however, because mold damage may occur.

Also, there are cases when double ejection is required to "kick" a reluctant part off the ejector pins. This means that the ejector system comes forward, returns, and comes forward again before finally returning in preparation for the next cycle. This practice is called *pulsing* and, in effect, it doubles the amount of ejection time and the amount of wear on the ejection system. It is not recommended.

Part-removal Time

In those cases where an operator (or robot) must be used to remove the part from the mold, time must be included in the cycle for that operation. This can usually be performed in 2 or 3 seconds, depending on the degree of difficulty. Even if an operator is not used and the parts fall automatically, time must be allowed for the parts to fall clear of the mold before it closes again.

Mold-inspection Time

This is a procedure often overlooked in estimating total cycle times. It is a good practice to have the operator (when one is present) look the mold over before closing the gate to start the next cycle. This takes only a second, but can save thousands of dollars if a part is stuck, or slides have been knocked out of adjustment, or other similar situations have occurred. During this time, the operator may quickly clean off the mold surfaces to remove flash, grease, outgassing, or other contaminants. And, finally, the operator may need to apply a mold release periodically. All of these operations must be allowed for in establishing the total cycle time.

DISTANCE

Control of distances is critical to producing high-quality products at reasonable cost. This is primarily because excessive distance requires excessive time, and, as stated earlier, time is money. Because distance is so closely related to time, the various functions involving distance are basically the same as those listed above for time.

Mold-close Distance

As noted, there are two phases to mold closing: the initial close, which covers the major portion of the closing distance, and the final close, which covers the small remaining portion of the closing distance (Figure 3-9).

The distance covered by the initial mold closing should take the mold halves to within 1/4 to 1/2 in. (0.6 to 1.3 cm) from touching. This closing

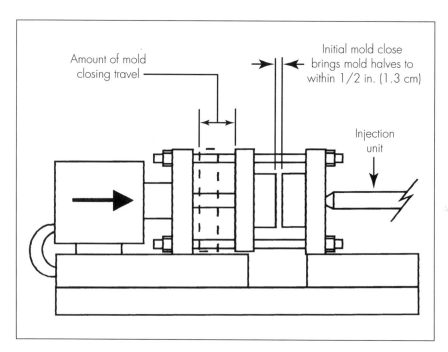

Figure 3-9. Mold-closing distances.

distance is normally traveled under high speed to get the mold closed as soon as possible so the overall cycle time can be minimized. But, if the mold halves are simply allowed to slam together under that high speed, they will eventually crack and break. Also, there may be an obstruction in the mold (such as a broken piece of plastic) which will cause damage if the mold is allowed to close up on it. So the mold closes quickly, but under very low pressure, only until the mold halves are close to touching. This distance is measured from the point at which the mold begins to close, and goes to the gap mentioned above.

At that point, the mold is slowed to a crawl. This occurs for the entire 1/4 to 1/2 in. distance of travel, until the mold halves are closed tightly against each other. This final closing is done slowly to minimize closing damage. If a foreign object is caught in the mold, the closing action will stop at this point and the object can be removed. If the clamp were closed under high pressure, any foreign object would be crushed, causing damage to the mold.

After traveling the final 1/4 to 1/2 in., the mold is fully closed under full clamp tonnage and the injection phase is allowed to begin.

Injection Distance

As mentioned earlier, the injection process is performed in at least two phases: initial injection and injection hold (Figure 3-10). Initial injection distance must be set to ensure that approximately 95 percent of the intended material is injected. This distance varies depending on how big the machine is and how much of the barrel capacity is being injected for one shot. As stated, the ideal shot size is 50 percent of the barrel capacity. So, if the machine is rated as having a 6-oz (170-g) barrel capacity (in styrene), the ideal single shot size is 3 oz (85 g) (50 percent). The limit switch governing that shot size would therefore be set halfway back on the measurement scale. This scale is usually physically located on the injection barrel, but sometimes is part of the electronic control system. In any case, the scale is incremental and can be adjusted anywhere between 0 and 100 percent of the barrel capacity.

Injection-hold Distance

After the initial injection setting allows 95 percent (approximately) of the required material to be injected, the machine switches to holding pressure. This finishes filling the mold and holds pressure against the material that was injected. The point at which the hold pressure takes over should be set almost at the very end of the stroke of the injection screw.

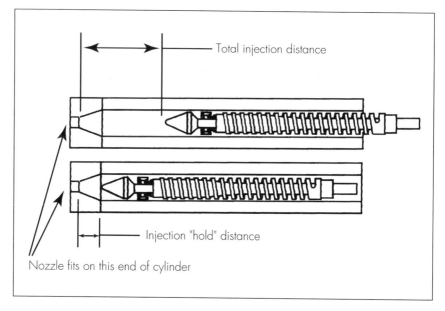

Figure 3-10. Injection and hold distances.

Cushion (Pad)

There should be a pad, or cushion, of material left in the barrel for this hold pressure to be applied against (Figure 3-11). The cushion should be approximately 1/8 to 1/4 in. (0.3 to 0.6 cm) thick.

The cushion is established by creating a total shot size that is slightly larger than that required to fill the mold. For example, if the amount of material required to fill the mold is 2.9 oz (82.2 g), the total shot size would be established at approximately 3.0 oz (85 g). This would then be adjusted (increased or decreased) during setup until the 1/8-in. cushion is developed by changing the set point for the screw return, which is discussed below.

The thickness of the cushion is critical. It must not be less than 1/8 in. because anything less is difficult to control accurately and there is a good chance that the cushion will go to zero on a random basis because of inconsistencies in the specific gravity of the melt. If the cushion does go to zero (or *bottoms out*), there will be no pressure against the material in the mold and the molded part may warp, crack, or simply not fill. Also, the shrinkage will vary and the part may not be dimensionally acceptable.

If the cushion is more than 1/4 in. thick, the plastic in the cushion might cool and begin to solidify because of all the steel surrounding it. This

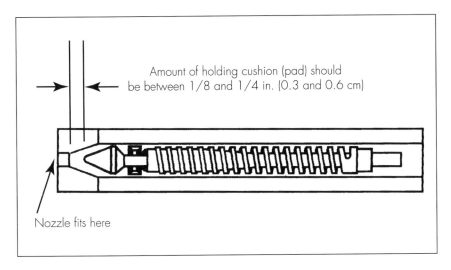

Figure 3-11. Hold cushion.

will cause a blocked nozzle, which will result in slow-flowing material or no flow at all.

Screw-return Distance

The primary function of the screw-return process (Figure 3-12) is to prepare the charge for the next shot. After the injection phases are completed, the screw turns to bring fresh material forward in the heating cylinder. As the material is brought forward, it pushes the screw backward while it is turning. This continues until the screw has returned to the set point, at which time it stops turning. It must be set at a point at which there is slightly more material in the barrel (for the next shot) than is required to fill the mold. What is extra will be used to establish the cushion. The rate of screw return can be altered by adjusting the revolutions per minute (rpm) of the screw drive motor. Generally speaking, the higher the rpm, the faster the screw returns. But, there is a range of rpm in which each material is best prepared for injection, and each screw design has an effect on proper screw rotation speed. The overall rpm range should normally fall within 30 to 160.

Mold-open Distance

To break the vacuum that was created in the cavity image during the injection process, the mold must be opened slowly. After the mold has opened

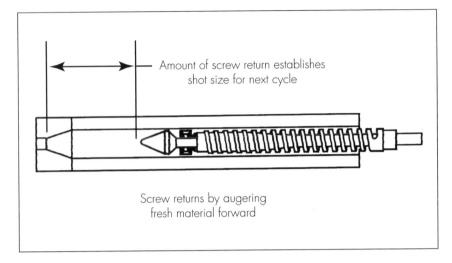

Amount of screw return establishes
shot size for next cycle

Screw returns by augering
fresh material forward

Figure 3-12. Screw return distance.

1/4 in. (0.6 cm) or so, the vacuum on the stationary side is relieved and
the mold can be allowed to open the rest of the way at a faster rate. The
faster rate is desired so that the cycle can be completed quickly and the
next cycle started.

If a mold contains "actions," such as slides or cams, there may be a
requirement to open the mold slowly for a longer distance. This might
range from the original 1/4 in. to 2 or 3 in. (5 to 8 cm) or more.

Once the mold has opened enough to break the vacuum (and far enough
to clear the actions), it may be allowed to open fully. The total distance a
mold is to open should be no more than absolutely necessary because
it takes time for this to happen and time is money. That being the case,
how much is necessary?

The mold should open a total distance equal to twice the depth of the
molded part. For example, if the part being molded is a square box with a
depth of 6 in. (15 cm), the mold should be allowed to open no more than
12 in. (30 cm). If possible, this dimension should be made smaller. There
needs to be only enough open space to allow the finished part to fall clear
of the mold after ejection, or for the operator to reach in and remove the
ejected part.

If the mold is running with an operator who must physically remove
the part from the ejector system, the mold-open distance should be increased to whatever is necessary from the standpoint of safety and com-
fort, in addition to simply allowing enough room for manipulating the
part. In most cases, this will not exceed 2 1/2 times the depth of the part.

Ejection Distance

The amount of ejection required is only that which will push the part free from the mold (Figure 3-13). The only area of the part that is ejected is that which is confined in the moving half (B half) of the mold. If that area has a maximum depth of 1 in. (2.5 cm), then theoretically only 1 in. of ejection is required. If more is used, it takes additional time, and time is money. If less is used, the part will probably not fall free and will get stuck, causing damage if the mold closes on it.

It is a good practice to measure how much ejection is required and then add 1/8 to 1/4 in. (0.3 to 0.6 cm) to make sure the part is well clear of the mold surface.

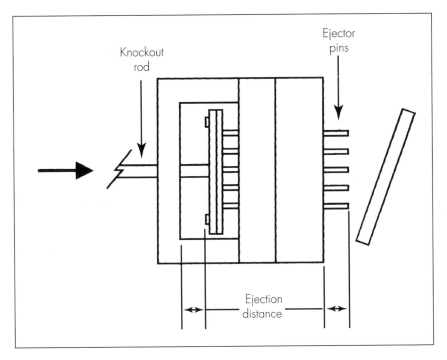

Figure 3-13. Ejection of finished part.

DETERMINING INJECTION-MOLDING COSTS

At some point in the product development and manufacturing process, it becomes necessary to establish the estimated or actual costs to produce a specific product.

What Information is Needed?

Injection molding is considered a primary manufacturing process. For that reason, this section will not address the details involved in computing secondary operations and/or packaging costs. We will look solely at injection-molding costs.

The following items are those needed for calculating actual injection-molding manufacturing costs:

1. Material costs.
      a. Raw material.
      b. Recycled material.
      c. Scrap allowance.
      d. Estimated regrind buildup.
2. Labor charges (if not included in standard machine rate).
      a. Straight time.
      b. Overtime.
3. Machine rate (hourly).
      a. Setup charges.
      b. Scrap allowance and downtime.
      c. Number of cavities in mold.
      d. Minimum number of cavities allowed.
      e. Cycle time per shot, in seconds.
4. Tooling charges (if amortized over product volumes).
      a. Initial mold costs.
      b. Maintenance costs.
      c. Volume for amortization calculations.

Material Costs

When we consider material costs, we must look at the options. First, can we use regrind? If so, what percentage? Then we must estimate how much regrind our process will generate on its own. To do this, we must estimate the volume of plastic needed to mold our parts and compare that to the volume of plastic needed to fill our runner system, assuming we are using a standard runner system. If a hot runner system (or insulated runner system) is being used, we do not need to consider the material needed to fill it. Let's look at the hypothetical situation in Figure 3-14. In this example, we will be molding a four-cavity mold, each cavity producing the same product. The standard runner system has been designed to normal specifications. The plastic material is polycarbonate, and the part has a nominal wall thickness of 0.075 in. (0.19 cm).

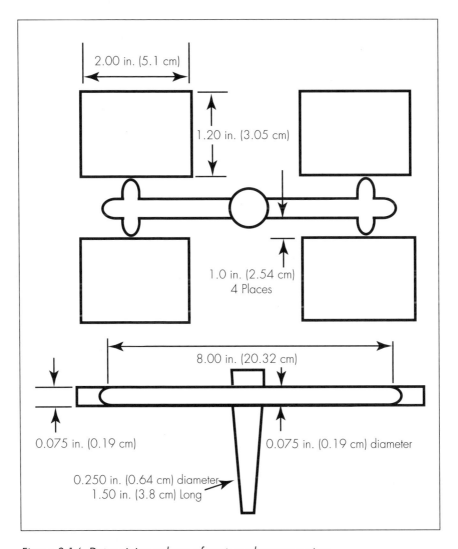

Figure 3-14. Determining volume of parts and runner system.

To determine material costs, use the following three-step formula:
1. Determine volume of part (cubic inches or cubic meters).
2. Determine weight per unit volume (specific gravity × 0.0361).
3. Determine cost per total volume of part (cost/lb × lb/in.^{3} × in.^{3} [cost/kg × kg/m^3 × m^3]).

First, we estimate the volume of the runner system. To do this, we take the cross-sectional area and multiply it by the total length of the runner.

The cross-sectional area is calculated by taking the radius of the 0.075-in. diameter runner and multiplying it by itself (r^2), then multiplying that figure by the constant, π (3.1416). Thus, 0.075/2 = 0.0375 × 0.0375 = 0.0014 × 3.1416. So, the area of the runner cross-section is 0.0044 in.² (0.0284 cm²). Now, we multiply that by the length of the runner (8.0 + 1 + 1 + 1 + 1 × 12 in. [30.5 cm]) to get 0.0530 in.³ (0.8685 cm³), which is the volume of plastic needed to fill the runner. Now, we calculate the sprue volume in the same way, but, because of the taper, we can divide the answer in half. In our example, the main sprue diameter is 0.25 in. (6.4 mm), and the length is 1.5 in. (38 mm). Performing the same type of calculation as above and dividing in half, we find the sprue volume to be 0.0368 in.³ (0.603 cm³). Adding this to the runner volume, we arrive at a grand total of 0.0898 in.³ (1.47 cm³) for the total runner system.

Next, we estimate the plastic needed to fill the parts. We take one part and multiply our answer by 4 (because it's a four-cavity mold) to get the total material required. In the case of the parts, we calculate volume by multiplying the 2.0-in. length by the 1.2-in. width to get 2.4 in.² (15.5 cm²). Multiplying that number by the wall thickness of 0.075 in., we get 0.18 in.³ (2.95 cm³) per part. Multiplying that by 4 (four cavities), we get a total of 0.72 in.³ (11.8 cm³) for all four parts.

So, for the runner we have 0.0898 in.³, and for the parts we have 0.72 in.³. Our total requirement for plastic material for a single cycle is found by adding these together to get 0.8098 in.³ (13.27 cm³). To determine how much weight of plastic will be needed, we multiply that number (0.8098) by 0.0361 (which is a standard conversion factor to lb) to get 0.0292 lb (0.013 kg). This number is multiplied by the specific gravity of polycarbonate (or whatever material we are using) which is approximately 1.2. Multiplying 0.0292 × 1.2, we get a weight of 0.03504 lb (0.016 kg) of plastic required for each production cycle we run. Now we take that total and multiply it by the cost per pound for polycarbonate (approximately $4.00 for purposes of our exercise) and find the cost per cycle to be $0.140. Each of the parts (4) would then require $0.035 worth of plastic to produce.

Use of regrind. In most cases, the use of regrind is acceptable at specific levels, and will reduce the overall cost of manufacturing the product. Regrind can usually be acquired at a per-pound cost of approximately 50 percent of virgin per-pound costs. While the specific gravity may be different, the total material cost of regrind can be determined using the same formulas as those shown above.

It may be possible to use 100-percent regrind in a specific product, because regrind that has not been abused will retain as much as 90 percent of the properties of virgin. Testing will help determine the acceptable level of regrind, but the typical average is approximately 15-percent regrind mixed with 85-percent virgin.

Impact of runner versus shot size. In calculating material costs, it is important to understand how much regrind is being generated by the products being molded. Then we can assess the use of that regrind in possibly reducing material costs.

The normal accepted level of regrind use is 15 percent. If our runner system is up to 15 percent of the total shot size, we can use our own generated regrind. If it is more than 15 percent, we may have to store the amount over 15 percent and use it elsewhere or sell it to other molders.

In the case discussed above, our total shot size contained 0.8098 in.^{3} (13.27 cm^3), while our runner had a total volume of 0.0898 in.^{3} (1.47 cm^3). Taking 15 percent of 0.8098 shows us that we can use up to 0.121 in.^{3} (1.98 cm^3) of regrind per shot. Our actual regrind volume of 0.0898 in.^{3} is much less than the allowed 0.121 so we can use all the regrind generated by this runner system. All we have to do is mix it back into the virgin that we will be using. In that way, the runner system is actually molded for "free" and does not have to be used in determining the total manufacturing costs for this product. You can see by this exercise that if we make sure our runner system is always less than 15 percent of the total shot size, it can be molded for free.

Labor Costs

Usually the cost of labor is included in the basic machine rate (see "Machine Costs," below), but if this is *not* the case, it can be easily calculated. To do so requires that a basic hourly labor rate be established, and then added to average burden, benefit, and overhead values, which usually add 100 percent to the hourly rate. For instance, if the average pay for a machine operator is $7 per hour, the overhead, burden, and benefit values will probably also be worth $7. This gives a total labor charge of approximately $14 per hour. This charge, divided by how many pieces are molded in an hour, will give a total amount to be added to the cost of each piece, for labor. Remember also that any overtime that might be planned in order to meet production schedules must be taken into account. This overtime rate must also be factored by the burden, benefit, and overhead values.

Machine Costs

A few things need to be addressed before the actual machine costs for injection molding a product can be determined. First, the size of machine must be identified; second, the geographical location of the machine must be known; and third, whether the labor cost of an operator is to be included in the machine rate must be decided. Also, we must calculate a

standard setup charge for placing the mold in the machine, getting it running for production, and removing it after the run is completed. Start with
estimating the size of machine required.

   Determining machine size. Determine the required machine size by
establishing two things: how much clamp tonnage is necessary, and how
big an injection unit is needed.

   Required clamp tonnage is determined by calculating the projected area
of the parts and runner being molded. In the case we are working with,
we multiply the length of a part (2.0 in.) times the width of the part (1.2
in.) to get a projected area of 2.4 in.^{2}. We have four cavities, so we must
add the area of all four parts. This gives us 2.4 × 4 = 9.6 in.^{2} (61.9 cm^2) of
projected area for the parts. Now we add the area for the runner. The width
of the runner is its diameter, 0.075 in., and the length is 12 in. Multiplying
these numbers gives us a total of 0.9 in.^{2} (5.8 cm^2). Now we add the area of
the runner to the area of the parts and arrive at a total of 10.5 in.^{2} (67.7 cm^2)
projected area. At this point, we must multiply the projected area (10.5
in.^{2}) by a factor representing the number of tons of clamp force. As mentioned, a rule of thumb is to use 5 as the factor, but it can be from 2 to 8,
depending on how easy the plastic flows when it is injected. Our material
of choice, polycarbonate, does have a factor of 5, so we multiply our projected area (10.5) by a factor of 5 tons (44.5 kN). This equals 52.5 tons (467.2
kN) and represents how much total clamp force will be required to hold
the mold closed during the injection process. Adding a safety factor of 10
percent to this number gives us a machine with at least 57.75 tons (514
kN) of clamp. Quite likely, we would place this mold in a 60-ton (534-kN)
machine. If we do not have the exact size machine available, we can place
it in a larger one, but *never* in a machine that exceeds 10 tons per square
inch (12 kN per square centimeter) of projected area, as that is enough to
collapse the steel mold and press the mold into the machine platens, dam-
aging the mold and machine. In our case, that would be a machine with 10
tons times our projected area of 10.5 in.^{2}, for a total of 105 tons (934.5 kN).
Our machine range then is 60 to 105 tons.

   Next, we want to determine the injection unit requirements. This, too,
is calculated using the projected area of the runner and parts. The number
above, 10.5 in.^{2}, is the one we will use.

   As discussed in Chapter 2, all plastic materials have a specific flow rate.
Some flow easily, which means they do not require high injection pressures, and some are more difficult to push, requiring higher injection pressures. An average value for injection pressure is 10,000 psi (68,940 kPa).
Because psi stands for pounds per square inch, we can see that injection
pressure is given as the force needed over a certain area, i.e., the projected
area we calculated for the parts and runner. So we must have a machine

capable of producing 10,000 psi injection pressure. Most machines are designed to provide at least 20,000 psi (137,890 kPa), so this should not be a problem.

We must now determine the exact amount of material to be injected during one cycle. We have already calculated (on page **58**) that to be 0.8098 in.^{3}, or 0.0292 lb of plastic required for each production cycle we run. Machines are rated as to how many ounces of material they are capable of injecting at one time, so we must convert our number to ounces by multiplying it by 16 (ounces per pound), which gives us 0.4672 oz, or just under 1/2 oz (14 g). An ideal situation allows us to inject half the barrel capacity every shot, so we would want to place our mold in a machine with a 0.934-oz (26.48-g barrel. Of course, we will look for a standard size, which would be 1 oz (28.35 g). Remember that our rule of thumb states that we can use anywhere from 20 to 80 percent of the barrel capacity per shot. In our case, that would allow us to use any machine with a barrel size of from 5/8-oz to 2 1/2-oz (17.72- to 70.88-g) capacity (1/2 oz is 80 percent of a 5/8-oz machine, and 20 percent of a 2 1/2-oz machine).

So, now we know that our four-cavity mold should be run in any machine with clamp tonnage from 60 to 105 tons, and with a barrel capacity of 5/8 to 2 1/2 oz of plastic material.

Determining machine location. Utility rates and operating costs are different in different parts of the country. In the United States, the Northeast and Far West have the highest, and the South Central states and Southeast have the lowest. This means that the cost of doing business changes with geographic location. The *machine-hour rate* (MHR) can be defined as the hourly costs involved for the operation of a machine, and includes such items as overhead, management salaries, and plant maintenance. The MHR should reflect these costs relative to where the machine operates. The average MHR value can be selected from Table III-4 according to machine size, then increased or decreased by a percentage that reflects the machine's location.

These are average costs and normally include one operator. For Far West and Northeast locations, add 25 to 50 percent, depending on local utility rates and wages. For South Central and Southeastern locations, deduct 15 to 25 percent, depending on the same factors. As the table indicates, it makes good sense to place a mold in the smallest machine possible, to keep the manufacturing costs low.

Calculating machine cost. Now that we have critical machine-hour rates, we can determine molding costs for our four-cavity mold.

We need to estimate an overall cycle for our production run. The cycle is based on many things (which are addressed in Chapter 4), but what we

Table III-4. Average Machine-hour Rates

Machine size, tons (kN)	MHR, dollars per hour
0-100 (0-890)	25-30
101-200 (891-1780)	30-35
201-300 (1781-2670)	35-45
301-500 (2671-4450)	45-55
501-750 (4451-6670)	55-75
751-1000 (6671-8900)	75-100

are interested in here is the total, or gate-to-gate, cycle. The main item of importance in determining cycle time is the time it takes to cool the plastic, and that is dependent on wall thickness. Table III-5 gives some average overall cycle times for various wall thicknesses. If the mold is a complicated one, such as an unscrewing mold, or a large one, the cycle may be much longer because of extended open and close portions of the cycle.

Table III-5. Typical Cycle Times Determined by Wall Thickness

Wall thickness, in. (mm)	Overall cycle time, seconds
0.060 (1.5)	18
0.075 (1.9)	22
0.100 (2.5)	28
0.125 (3.2)	36

Now that we have a guide for estimating cycles, we can estimate the number of pieces produced per hour. This is done by dividing the number of seconds in an hour (3600) by the gate-to-gate cycle for our product, and multiplying the result by the number of cavities being molded at one time. The formula is:

(3600/cycle) × number of cavities = pieces per hour

In our case, the cycle-time chart shows an average cycle for a part with 0.075-in. wall thickness to be 22 seconds. We divide 3600 by 22 seconds to get 163.63. This we multiply by the number of cavities we have (four) to get a total of 654.52 pieces produced in an hour's time. If we run our mold in a 100-ton machine, the hourly rate for that size machine is $30 (from the MHR table). So we know that it would cost $30 to produce 654.52 molded parts. Dividing the $30 by the number of parts (654.52) gives us a cost of $0.046 per piece. At this time we should add a scrap factor, usually 10 percent to the value of the parts. For us that would be $0.005. Adding that to our cost, we arrive at a molding cost of $0.051 each.

Should operator cost be included? There are pros and cons to the issue of operator cost. Each case should be studied and assessed on its own merits. If a molding facility has a set number of operators (such as might be the case in a captive shop) and does not wish to fluctuate that number, then it would be wise to include operator cost in every quote. On the other hand, if the facility has seasonal requirements and/or uses robots for most of the molding machines, then the cost of an operator should not be included. This may give the company an edge in obtaining new business.

If a facility is not labor-intensive, it must use automated equipment or sophisticated tooling to achieve the necessary production requirements. By doing so, it incurs extra costs that must be included in determining the manufacturing cost of any product.

So, in either case (using an operator or not), there are costs involved that usually cancel each other out and do not greatly affect the actual machine-hour rate.

Estimating setup charges. A setup charge is a one-time fee assessed each time a mold is placed in a press for a production run. Normally it is derived from establishing the amount of time required to make the setup and multiplying that by the hourly rate for that machine. Thus a 2-hour setup on a 300-ton (2670-kN) machine would be approximately $90 (2 hours × $45). The normal time to complete an average setup is 2 to 3 hours.

For companies that have high volumes of production, or leave a specific mold in a machine for long periods of time, the cost of setup for that mold is usually absorbed. But in cases where volumes are smaller, or molds are changed frequently (as in most custom-molding facilities), a setup charge is usually levied. This charge could run from as little as $50 to as much as $500 or more, and is usually invoiced separately from the cost of molding the products. The decision to charge or not is entirely up to management, which must consider such things as customer loyalty and level of capacity at the time. In some cases, especially for low-volume runs, the setup charge is amortized over the total number of pieces being produced. In other cases, it is absorbed in overhead and considered a standard part of doing business.

Tooling Costs

Sometimes tooling costs are not included in the manufacturing cost estimates. They may be paid for by a special fund dedicated to total tooling costs for a specific plant site. Regardless of how the tooling is funded, there is a definite cost associated with building molds and secondary tools for a specific product design.

Standard practice. In most cases, tooling (mold) charges are not amortized over a specific volume. This is because volumes never stay fixed and are continuously changing as market influences change. To keep from having to constantly adjust manufacturing costs for these changes, molders treat the cost of tooling as a separate issue. Usually the billing practice for tooling costs requires that the customer pay a third of the cost as a down payment to the moldmaker, a third on delivery of the mold to the molder, and the final third within 30 days of tool acceptance by the customer. This practice keeps the moldmaker from being burdened with funding the initial building of the mold and risking loss of capital, which can run into hundreds of thousands of dollars.

Amortizing tool costs. If tooling costs must be amortized, for whatever reason, it is usually done over the first year's production run, which has been negotiated and guaranteed by the customer. That way, even if the parts are never produced, the molder and moldmaker are paid for their involvement. This practice results in unusually high costs for the first year's molded parts, with each part absorbing a portion of the total tooling costs. If the run is only a few hundred pieces, each piece may become so costly it cannot be sold. This is the primary concern about amortizing tool costs.

Maintenance costs. Whether tooling costs are amortized or paid for up front, there is a need to address everyday maintenance charges for upkeep of the mold. As a rule, maintenance will result in an annual charge that is equal to approximately 8 percent of the original tool cost. For example, if a mold costs $50,000 to build, we can assume that, every year, we will spend approximately $4000 to maintain that mold. This money is used for such things as waterline hoses, ejector pins, lubricant, damage repair, and other items required to keep the mold in top condition. This money can be included in standard overhead charges, added to the original cost of the mold, or billed as the items are activated. If the volume of production is great enough, the charges can be added to the piece price (amortized) and absorbed by the molder as required.

Adding It All Up

At this point, we have enough information to determine the actual costs involved to mold our product. All we need do is add the material costs

($0.035) to the molding cost ($0.051 each) to arrive at $0.086 each. If there are any secondary operations to be performed, the costs would be added to this figure. Secondary operation costs are determined in the same manner as molding costs, by determining pieces per hour and any material costs. In addition, packaging, inspection, and freight charges may have to be added. But the primary manufacturing cost has been determined to be $0.086. This number can also be used as a "sanity check" to determine whether a vendor is charging a reasonable price for molding a product for you, the customer, or to develop a target cost for in-house molding operations. Remember, though, that no profit has been included in this number.

SUMMARY

Although there are more than 200 recognized parameters that affect the injection molding process, those of primary importance can be categorized within four major divisions: temperature, pressure, time, and distance. An adjustment to any one of these has a direct influence on some, or all, of the other parameters.

For minimum-defect, high-quality production, it is critical to control as many parameters as possible. The more parameters that are controlled, the higher the quality level of the products being molded.

Pressure is the one parameter that shows an immediate response to an adjustment. All others require varying degrees of time to show results.

There are two basic styles of clamp mechanisms: hydraulic and mechanical. In addition, all-electric machines have been introduced that are a variation of mechanical systems.

QUESTIONS

1. What are the four groups into which all primary parameters are categorized?
2. What two methods are utilized for heating plastic in the injection barrel?
3. How can energy costs for heating the barrel be reduced by 25 percent?
4. What is the total length of time required until a molded part is totally cooled?

(a) 2 hours	(b) 1 day
(c) 1 week	(d) 1 month

5. What is the difference between *injection pressure* and *hold pressure*?
6. In your own words, how would you define *back pressure*?

7. List one advantage each for the hydraulic and mechanical clamp systems.
8. What is meant by the term *gate-to-gate cycle*?
9. Why should the mold open slowly for at least 1/4 in. (0.6 cm)?
10. Why is control of distance so critical to producing parts at low cost?
11. What information is needed for determining molding costs?
12. What is meant by *machine-hour rate*?
13. What is meant by *amortized* tooling costs?
14. What is meant by the term *setup charges*?
15. What is the *average annual maintenance cost* for molds?

Optimizing the
Molding Parameters

4

THE NEED FOR CONTROL

To take advantage of the numerous benefits of injection molding, it is necessary to control as many facets of the process as possible. In Chapter 3, we showed how each of the multitude of parameters that affect processing can be placed in one of only four categories. Actually controlling those parameters can be just as simple. But is it really necessary? Should valuable resources (time, money, personnel, space) be allocated to this end? The answer to both questions should be a resounding "Yes!" The reason for this is that any successful method for controlling the quality and the cost of a product depends heavily on the consistency of the process used to manufacture it. Consistency can be achieved only by tightly controlling as many parameters as possible during the manufacturing process. This does not mean that adjustments cannot be made once the job is running. It simply means that proper control allows for accurate, meaningful adjustments, when they are necessary. If proper control of parameters is attained, consistency follows. This consistency takes the form of part quality and part cost.

Part Quality

Part quality requirements are usually determined through discussions and agreements between the owner of the product design and the manufacturer of the product. This may be the same person (or group), or may be two individual entities, although in some cases the owner of the product design sets down the requirements without any discussion with anyone else. In any case, a set of requirements is established that determines how the finished product will respond to a given set of circumstances. These requirements may include mechanical, physical, aesthetic, thermal, and other values, and tolerances are placed on each of these values, if practical. For example, the product designer may wish to have a product that withstands high temperature exposure for extended periods of time. Or there may be a requirement that states the product should be a specific length. Both of these requirements must have a value placed on them. The temperature requirement might be listed as 160° F (71° C), and the length requirement might be 10 in. (25 cm). Reasonable tolerances are placed on these values and the manufacturer (molder) then understands the relative importance of those two requirements. The 160° F becomes 160° F ±10° (71° C ±5.6°) and the 10 in. becomes 10 ± 1/2 in. (25 ±1.3 cm).

Understanding the importance of the requirements, the molder knows how much control is essential to achieve the necessary level of consistency to meet these requirements at the most efficient and economical level of manufacturing. The molder knows, for instance, that the 10-in. length dimension can be made slightly larger if high injection pressures are used, or slightly smaller if low injection pressures are used. But the molder must be aware that the closer that dimension must be held, the closer the control must be maintained on the pressure parameter. This does not say that all parts must be molded to extremely high quality levels. It simply states that once the required level of quality is understood, it can be met and maintained through proper control of molding parameters. The lower the level of quality required, the easier it is to control the processing parameters.

Part Cost

The cost to mold a part is determined in a variety of ways, as discussed in Chapter 3. Once that cost is estimated (and successfully quoted), it becomes the responsibility of the molder to ensure that the cost is not exceeded. Actually, the molder should try to reduce the true cost if possible and this reduction can be passed on to the customer, kept as additional profit by the molder, or shared equitably by both. Proper control of the process parameters will allow that cost maintenance (or reduction) to occur. Maintaining a consistent cycle, with consistent parameters, will practically guarantee a zero-defect product.

It must be stated here that every part molded, whether good or bad, is bought by someone. The customer buys the good parts, and the molder buys the bad parts. So it just makes good sense for the molder to strive for a zero-defect situation, and to keep the manufacturing costs to a minimum. This holds true whether the molder is making inexpensive flowerpots, or precision-molded electronic devices. There is no good reason for the molder to produce bad parts that the customer will not buy.

So, it is necessary to produce consistently acceptable products. This can be done by maintaining consistent processing cycles through consistent parameter control. Let's look at some of the methods used for doing that.

Parameter Effects

Table IV-1 shows some of the property values that can be adjusted by a plus or minus change in some of the more common molding parameters.

Table IV-1. Molding Parameter Adjustments and Effects

Parameter	Property effect
Injection pressure (+)	Less shrinkage, higher gloss, less warp, harder to eject
Injection pressure (−)	More shrinkage, less gloss, more warp, easier to eject
Back pressure (+)	Higher density, more degradation, fewer voids
Back pressure (−)	Lower density, less degradation, more voids
Melt temperature (+)	Faster flow, more degradation, more brittle, more flashing
Melt temperature (−)	Slower flow, less degradation, less brittle, less flashing
Mold temperature (+)	Longer cycle, higher gloss, less warp, less shrinkage
Mold temperature (−)	Faster cycle, lower gloss, greater warp, higher shrinkage

These are just some examples. But notice how some properties are changed in the same way by different parameters. For instance, less shrinkage can be attained by either increasing injection pressure or increasing mold temperature, and less degradation can be achieved by lowering back pressure as well as lowering melt temperature. These examples are shown to demonstrate that the basic molding parameters do work closely together, and that changing a parameter in one area may affect a value of some property in another area. By understanding this relationship, it is possible to minimize the number of adjustments when it is necessary to make a correction due to an unexpected change in some variable of the process.

WHAT ARE THE PROPER PARAMETER VALUES?

In other words, what's the best setting for the injection pressure, back pressure, melt temperature, mold temperature, etc.? It all depends on the material being molded and the type of mold being used, as well as the status of the injection machine. The parameters of time and distance were discussed in adequate detail in Chapter 3, but we include some aspects of them here. We focus on the pressure and temperature parameters in this chapter. First, there are a few rules of thumb that apply.

The Setup Sheet

At the outset, it is necessary to understand that, in most molding facilities, it is common to have a setup sheet that lists a variety of the common parameters and the value of each. This is used to start up a mold at the beginning of production. A typical setup sheet format is shown in Chart 4-1. Notice in the areas identified as *Temperature, Timer settings, Pressure settings,* and *Miscellaneous,* there are two columns for listing values. The first column is marked *start-up,* and the second is marked *actual.* This is an accurate and appropriate type of form to use because it acknowledges that both start-up settings and other settings should be used for actual long-term production.

The reason that both sets of values are needed is that the parameters set for the initial parts will begin affecting each other immediately after starting up. They will eventually stabilize (after approximately 6 to 8 hours), and at that point they will be affecting the molding material in a different way from the initial start-up effects. For instance, if the barrel rear temperature is set to 475° F (246° C) at the beginning of a run, it takes approximately 45 minutes for the material inside that barrel zone to get close to that setting. Then the temperature controller begins to cycle off and on in order to maintain the temperature properly. This results in a "soaking" of the barrel until the entire zone is at the right temperature. In the meantime, the material going through that zone has been heated in different stages, with the first mass being heated to approximately 450° F (232° C), the next few batches being heated to approximately 460° F (238° C), and the final batches being heated to approximately 470° F (243° C) before the next and continuing material batches are heated to the correct temperature setting. (This holds true even if the material is allowed to soak for 30 minutes before beginning the run. When the run starts, all the parameters begin to stabilize.)

While all of this is going on, the major parameters would have been adjusted to accommodate the original batches of material, so items such as injection pressure would be set high enough to inject that material into the mold. After a few minutes, when the next few batches come through (at a higher temperature), they flow easier and tend to begin flashing if the injection pressure is still held at the higher setting. And, finally, when the properly heated batches come through, they are flowing so easily that they will, in fact, flash. Then, injection pressures must be adjusted downward to accommodate the hotter material. So by the time everything stabilizes, which may be 6 hours or more, most of the parameters have to be reset to different values from those at which they started.

This explains the reason for maintaining two separate parameter sheets or, at least, two separate columns of data on a single parameter sheet.

ABC MOLDING COMPANY   **SETUP SHEET**   **JOB#** _____

Customer Name:_____ Part Number: _____ Date: _____

Machine Data		Timer Settings		
Preferred Machine: _____		□ Automatic □ Semiautomatic □ Manual		
Clamp Tonnage Required: _____		**Area**	**Start-up**	**Actual**
Clamp Stroke: _____		Mold Open:	_____	_____
Misc. Equipment Req'd: _____		Injection Delay: _____		_____
_____		Injection Fwd:	_____	_____
_____		Injection 1st:	_____	_____
_____		Injection 2nd:	_____	_____
_____		Injection Hold:	_____	_____
		Mold Close:	_____	_____
Mold Data		Decompress:	_____	_____
Mold Number: _____		Ejectors Fwd:	_____	_____
Part Description: _____		Air Blow Off:	_____	_____
No. of Cavities: _____				
Shut Height: _____		**Pressure Settings**		
Mold Open Distance: _____		**Area**	**Start-up**	**Actual**
Runner Type: _____		Clamp:	_____	_____
Nozzle Type: _____		Accumulator:	_____	_____
Ejection Stroke: _____		Back Pressure:	_____	_____
Special Requirements: _____		Injection 1st:	_____	_____
_____		Injection 2nd:	_____	_____
_____		Injection Hold:	_____	_____
_____		Ejection:	_____	_____

Temperature			Miscellaneous		
Area	**Start-up**	**Actual**	**Area**	**Start-up**	**Actual**
Feed Throat:	_____	_____	Overall Cycle:	_____	_____
Barrel Rear:	_____	_____	Cushion (in.):	_____	_____
Barrel Center:	_____	_____	Screw RPM:	_____	_____
Barrel Front:	_____	_____	Clamp Speed:	_____	_____
Nozzle:	_____	_____	Ejector Speed:	_____	_____
Mold "A" Half:	_____	_____		_____	_____
Mold "B" Half:	_____	_____		_____	_____
Hot Runner Zone Settings:				_____	_____
1:____ 2:____ 3:____ 4:____				_____	_____
5:____ 6:____ 7:____ 8:____				_____	_____
9:____ 10:____ 11:____ 12:____				_____	_____

Prepared by: _____

Material Description: _____   **Date:** _____

Material Number: _____   **Approved by:** _____

Total Shot Weight: _____   **Date:** _____

Chart 4-1. Typical setup sheet. (Courtesy Texas Plastic Technologies)

Installing and Setting Up the Mold

Sizing and Inspection

A machine must be selected that is properly sized for the specific mold being installed. This is discussed later in the sections "How Much Clamp Pressure Is Required?" and "Determining Shot Size." After the machine is selected, it must be inspected to determine its status. Inspection items include:

- Proper hydraulic oil level;
- Heater bands in place and operating;
- Mold temperature controllers operable;
- Injection cylinder empty and screw forward;
- Hopper shutoff closed and hopper wiped clean;
- Proper material available and dried;
- Granulator clean and available;
- Safety gates and mechanisms operating and in good condition;
- Vent hoods clean and operating;
- Heat exchanger clean and operating;
- Machine lubricated, or autolubrication working and filled;
- Alarms and lights operable.

Installation Procedure

After the machine inspection is completed, the mold can be installed. The following steps should be taken, but although they are generic in nature, they do not pre-empt the machine manufacturer's instructions, which always take precedence.

1. Make sure that the mold has a connecting strap properly installed. This strap should connect the two halves of the mold so that they do not come apart during transportation. Normally this is a metal strap mounted across the A and B plate parting line. It is not a safe or proper practice to install the mold as two separated halves.
2. Start the machine, make sure the injection sled is in the full back position, and set the barrel heaters to the proper temperatures. The profile should run from a cool rear setting to a progressively hotter front zone and nozzle as outlined on the setup sheet. Turn on the feed throat cooling water.
3. Open the clamp wide enough to accept the mold. This is normally a dimension that equals a minimum of twice the height of the mold. This may require resetting the mold-open limit switches or control settings. Refer to the machine manual for instructions.
4. Lower the mold from the top of the machine (or slide it in from the side) using a chain fall or similar device, and bring the mold up against the stationary platen, by hand. The mold should rest against

the platen without assistance. This is accomplished by adjusting the location of the chain fall toward the platen. It is good practice to place a thick metal plate across the lower tie bars at this point. The plate will act as a safety catch in case the chain fall breaks or the connecting hook opens.

5. At this point, the mold must be raised and lowered slightly in an attempt to position it so the locating ring on the mold will slip into the locating hole on the platen. The chain fall should be connected so that the mold tilts slightly at the top (see Figure 4-1).

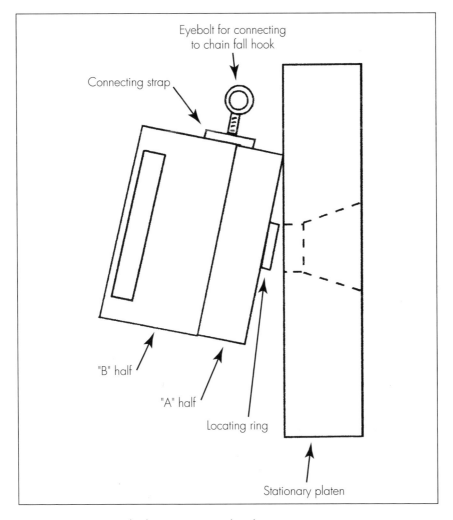

Figure 4-1. Inserting the locating ring in the platen.

6. The tilted mold should be placed slightly above the locating hole of the stationary platen and held against that platen as the mold is slowly lowered. The locating ring of the mold will automatically slip into the locating hole of the platen as the mold is gently lowered. A level can be placed across the top of the mold to assist in aligning the mold so it fits squarely on the platen. After leveling, the A half of the mold is ready to clamp in place.

7. Locate clamps, adjust them, and bolt the A half of the mold to the stationary platen. The mold should be mounted with at least one clamp in each of the four corners. If the mold is very wide, additional clamps should be placed along the long dimension. If the mold is very small, it may be possible to use only two clamps per mold half, although this is not recommended, and smaller, specially built clamps may be needed in order to use four clamps per mold half. *Note*: The mold-holding clamps are designed especially to focus clamping force against the platens. It is critical that they be properly aligned for this to happen. Figure 4-2 illustrates the correct procedure.

The ideal situation is to have the clamp placed exactly parallel to the platen. However, notice in the drawing that the clamp is *not* parallel to the platen. This is because it is difficult to adjust the clamp so that it is *exactly* parallel. So, the clamp will end up being tilted one way or the other. If the clamp is tilted with the heel closer to the platen than the toe, the clamping force will be directed away from the platen. That will result in very little clamping force actually directed toward holding the mold in place. The clamp will eventually break loose and the mold will fall. Therefore, the best thing to do is purposely adjust the heel of the clamp away from the platen. This directs the clamping force toward the platen through the toe of the clamp. The angle of direction should be minimal and can be such that the clamp heel is only 1/8 to 1/4 in. (0.3 to 0.6 cm) away from parallel. Note also that the clamp is slotted for linear adjustment and the clamping bolt is located close to the mold. The adjustment allows the clamping bolt to be located as close as possible to the mold because that too aids in creating maximum clamping pressure on the mold itself.

8. If ejector rods are required, they should be placed in the mold at this point. Slowly bring the clamp unit forward, under low pressure, to prepare for clamping the B half of the mold. This may require adjusting limit switches or settings. Check the machine manual for this information. Bring the moving platen up to within 1/4 to 1/2

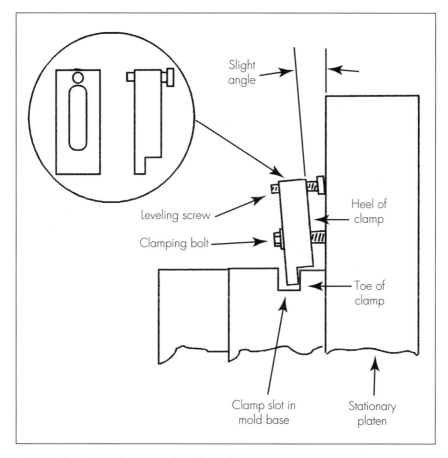

Figure 4-2. Proper alignment of holding clamps.

in. (0.6 to 1.3 cm) of the mold base and set limits for the high-pressure close to activate at that point. Continue moving the clamp unit forward until it touches the mold base. Allow the press to build up clamp pressure to the desired setting. This ensures that the mold is fully closed.

9. Shut off the machine. Locate clamps, adjust, and bolt the B half of the mold to the moving platen, making sure to follow the procedure mentioned in item 8.

10. Remove the chain fall hook, eyebolt, and connecting strap from the mold. To avoid losing the connecting strap, keep it mounted but swung out of the way and tightened down so it will not come loose and cause mold damage.

11. Recheck all clamps on both halves to make sure they are all tight. Start the machine and *slowly* jog the clamp unit open under low pressure, watching for any indication of the mold halves seizing or binding together. Open the mold approximately 1/2 in. (1.3 cm) and stop. Turn off the machine and fully check the mold to make sure it is properly mounted.

12. Start the machine and continue to open the mold slowly until the B half disengages fully from the A half. Then stop the mold at the point described for fully open. This would normally be a minimum of approximately 2 times the depth of the part being molded, to make sure the part will fall free after ejection. It is acceptable to open the mold farther, but it should not open farther than necessary because of the additional time required to do so. If there are slides or other actions in the mold, make sure they are still properly engaged on full opening if at all possible. This will minimize the potential for breakage. Check for broken springs or other obvious damage.

13. Adjust settings for proper ejection. Ejection should *not* pulsate. One stroke should be adequate for part removal. If this is not enough, there is something wrong and it should be corrected before continuing production. The amount of ejection stroke should not exceed 2 1/2 times the depth of the part in the B half of the mold (assuming ejection is located on the B half). Ejection stroke should be kept to a minimum and is only required to get that part of the plastic that is molded in the B half freely out of the mold. More than that only adds to the overall cycle time.

14. Lubricate all moving components such as ejector guides, leader pins, and slides. Wipe off all excess. Gently clean the cavity surfaces. Close the mold and turn off the machine.

15. Attach the hose lines from the mold temperature control unit. Blow air through the cooling lines of the mold to make sure they are not obstructed and to observe the proper path for connecting hoses. Do not loop the A half and B half together on a single line, but attach separate in and out lines for each half, and use separate control units for each half. Make sure there are no kinks in the hoses and that they will not be crushed or stretched when the mold closes or opens. After inspecting for proper attachment, activate the temperature control units and adjust for the proper temperature setting.

16. Recheck all clamps.

17. Check to determine if the barrel is up to heat. It normally takes 45 minutes to an hour for the barrel to properly come to preset tem-

peratures and soak. Make sure all heater bands are operable and properly connected.

18. Ensure that the hopper feed gate is closed and the hopper magnet is in position. Place fresh, dry material in the hopper. A purging compound may be required first, depending on what material was in the barrel last.

19. After the barrel is up to heat and has soaked for 10 to 15 minutes, open the feed gate on the hopper and allow material to drop through the feed throat and into the cylinder.

20. Purge the machine as follows.

    a. The injection screw should have been left in the forward position of the barrel when the last job was shut down. It should stay in that position while material is prepared for air shots. Activate the screw rotation until fresh material is brought to the front of the barrel. This will be obvious because the screw will spin freely at first, but will slow down considerably as the fresh material is brought forward.

    b. Set the screw return limits to the desired point and allow the screw to return to that point. The screw rotation will stop once the screw returns to the set point. Allow the material that was brought forward enough time to absorb heat from the cylinder. This will normally take only a minute or two.

    c. With the sled still in the back position, take three air shots. An air shot consists of injecting a full shot of material into the air, under molding pressure, and allowing it to accumulate on a special plate designed to catch purgings. Make sure that proper time is allocated between these air shots to allow the upcoming material to come to proper heat. This time usually amounts to the total cycle time of the job that will be running in production. Using a fast-acting pyrometer with a probe, measure the melt temperature of the material injected during the air shots. This is the temperature that must be controlled for proper molding. Adjust settings as necessary. If a different material or color is being used, 15 or 20 air shots may be needed to clear the old material out.

21. Set all limits for injection and cycle. These include injection forward speed and pressure, holding pressure, cushion distance, cooling time, mold-open and -close settings, and others as outlined in the following sections.

22. Prepare a full charge of material for the first shot. Bring the injection sled forward until the nozzle seats against the sprue bushing of the closed mold. The mold must be closed to absorb the force of the

injection sled against the sprue bushing. If the mold is open at this point, the A half will be pushed off the platen. Lock the sled control in place.

23. Open the mold and bring the clamp unit to the full open position.
24. Set the cycle indicator to manual, semiautomatic, or automatic, depending on requirements.
25. Close the safety gate to initiate the first cycle.
26. Observe the injection process. The pressures should be set so that a short shot will be taken first. Then pressure and time settings can be adjusted until a properly molded part is produced. This should be done over a long period of time (15 to 20 shots) and not hurried.

OPTIMIZING TEMPERATURE

Injection Cylinder Feed Throat

The first area of concern is at the rear of the barrel, in the feed throat section of the injection machine. This is the area directly under the hopper and is the first contact the plastic pellets have with the heating area of the molding machine. This is where pellets are dropped from the hopper in preparation for travel through the heating cylinder (barrel). A common problem that occurs here is *bridging* (Figure 4-3). This term refers to a condition in which too high a temperature in the feed throat causes incoming pellets to soften, or begin to melt, too early. They tend to stick together and do not fall freely through the feed throat. This results in a plug of sticky plastic blocking the feed throat and not allowing fresh material to fall through from the hopper. The injection barrel becomes starved for material and the machine cannot mold any product.

The temperature of the feed throat is controlled by a flow of coolant (usually water) and should be maintained in the range between 80 and 120° F (27 and 49° C). Remembering that this is the first temperature excursion the pellets are exposed to, it is better to maintain the feed throat as close to 100° F (38° C) as possible. This prepares the pellets for higher temperatures without shocking them as soon as they enter the feed throat.

Bridging is a dangerous condition and must be avoided. When bridging occurs, the injection barrel is starved for material, but there is still a certain amount of molten plastic residing in the heating cylinder. The longer it stays there, the hotter it becomes. After a few minutes, it degrades, giving off fumes and gases that are usually toxic. These gases build up pressure (because they are trapped and cannot get out of the barrel) until they explode backward through the feed throat and up through the hopper.

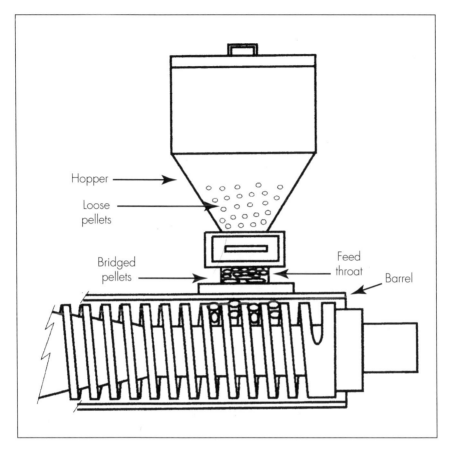

Figure 4-3. Bridging of the feed throat.

When this happens, the gases drag along some of the molten plastic from the barrel and spray this all over the area immediately surrounding the injection machine. This can cause severe burning to anyone in that area, and the toxic gases, if inhaled, can cause brutal damage to lungs and respiratory tracts. In addition, the hopper cover can be blown off, resulting in physical damage to equipment, as well as injury to personnel. Maintaining proper feed throat temperature will minimize the risk of bridging.

Injection Cylinder Nozzle Zone

The nozzle temperature should be controlled so that it is the same as (or 10° F [5.6° C] higher than) the recommended melt temperature of the material being processed. This information is shown generically in Chapter 3 for some of the more common materials (Table III-1, "Suggested Melt Temperatures for Various Plastics"), and specific values can be obtained from the material supplier. This melt temperature is the recommended temperature of the plastic as it leaves the nozzle of the injection machine and enters the mold. Taking a popular material like polycarbonate, we find that the suggested melt temperature from Table III-1 is 550° F (288° C). The nozzle temperature should be set at that same temperature, or even 10° F (5.6° C) higher, if the rate of throughput is high (more than 50 percent of the barrel capacity). The nozzle temperature is controlled by an independent temperature controller with perhaps only a single heater band, depending on the length of the nozzle (Figure 4-4).

Checking Temperature of Melt

Even though there is a temperature control unit that monitors the nozzle temperature, it must be noted that the actual temperature of the melted plastic can be checked properly only by using a probe-style pyrometer and thrusting it into the center of a purging shot. This is done by moving the injection sled backward, purging a full shot of plastic onto the purge plate of the machine, and immediately checking the temperature of that material with the pyrometer (Figure 4-5). This is the actual melt temperature.

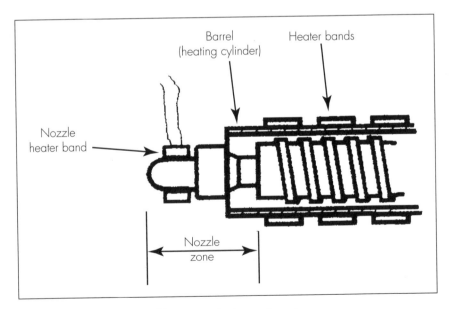

Figure 4-4. Nozzle zone of heating cylinder, or barrel.

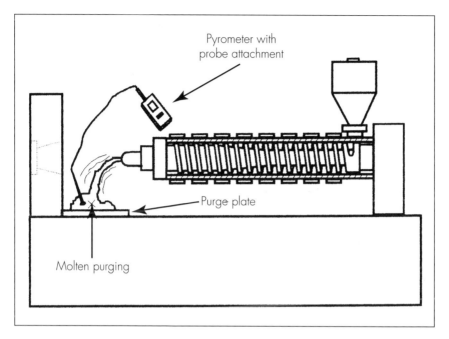

Figure 4-5. Checking melt temperature.

Injection Cylinder Front Zone

The front zone of the heating cylinder is located directly behind the nozzle and consists of the first third of the total length of the heating cylinder (Figure 4-6). The proper initial setting of temperature for this zone is 10 to 20° F (5.6 to 11° C) less than the nozzle setting. For the polycarbonate example, this would be 530 to 540° F (277 to 282° C). This zone is controlled by a separate temperature controller which is connected to a series of heater bands, usually three but up to six, depending on the total length of the barrel.

Injection Cylinder Center Zone

The center zone is a transition zone. It is located between the front zone and the rear zone and usually consists of the middle third of the total length of the barrel (Figure 4-7). The temperature for this zone should be set at the average of the front zone and the rear zone. The average for polycarbonate is calculated to be 500° F (260° C). This can be determined only after calculating the temperature of the rear zone so that an average can be found between the rear and front zones.

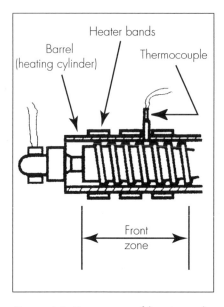

Figure 4-6. Front zone of heating cylinder, or barrel.

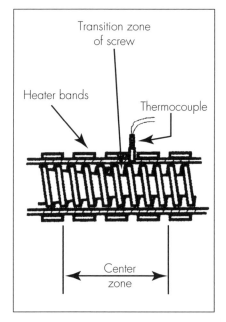

Figure 4-7. Center zone of heating cylinder, or barrel.

Injection Cylinder Rear Zone

From the feed throat, the plastic pellets are brought forward by the augering action of the injection screw until they enter the rear zone of the heating cylinder (Figure 4-8). This is where the material must be brought up to a temperature where it can become soft enough for the screw to homogenize (blend) it.

The proper temperature for this zone depends on the material being processed. If it is too high, the material may transfer heat backward to material in the feed throat, and this could cause bridging. Also, if it is too high, the material may become degraded before it leaves the heating cylinder to be injected into the mold cavity.

The proper rear zone temperature can be established by setting it at 15 percent less than the front zone temperature. In the polycarbonate example we are using, that temperature is 459° F (237° C) (15 percent less than 540), and we round it up to 460° F (238° C).

Now we can also determine the proper setting for the center zone. It should be an average of the front and rear. So, by adding the front and rear zone temperatures we arrive at 1000° F (538° F + 460° F). (The average is 500 [1000/2].) The center zone would be set at 500° F (260° C).

Injection Cylinder Summary

The initial injection cylinder temperature settings have now been established for our polycarbonate example. They are 550 to 560° F for the nozzle zone, 530 to 540° F for the front zone, 500° F for the center zone, and 460° F for the rear zone (288 to 293° C, 277 to 282° C, 260° C, and 238° C, respectively).

These settings are based on average residence times for the material. If the residence time is longer than average (i.e., shot size equals 20 percent of barrel capacity), the temperatures must be adjusted downward in 10 to 20° F (5.6 to 11° C) increments to minimize degradation. If the residence time is shorter than average (i.e., shot size equals 80 percent of injection barrel capacity), the temperatures may need to be adjusted upward in 10 to 20° F increments to increase the flowability of the plastic.

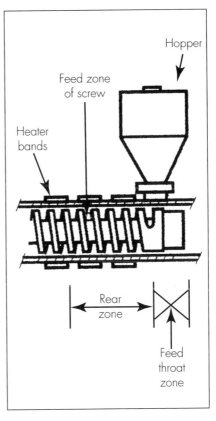

Insulation Jackets

Because tight control of all parameters is so important, insulation jackets should be placed around the injection barrel. These jackets are similar in concept to water-heater jackets. They trap the heat generated

Figure 4-8. Rear zone of heating cylinder. Molding process quality depends on critical temperature control in this zone of the injection barrel.

by the heater bands surrounding the barrel and force that heat back into the barrel instead of allowing it to escape to the atmosphere. This results in more efficient heater band utilization and will cut electrical costs by approximately 25 percent.

Preheating Material

It is good practice to preheat the plastic before it is placed in the machine hopper. Usually this is done at the time of drying the material (see "The Importance of Drying Materials" in this chapter). Preheating helps minimize thermal shock to the material when it is exposed to the cylinder heating process. For most materials, this preheat temperature will be somewhere between 150 and 200° F (66 and 93° C), depending on heat sensitivity. The exact temperature can be found in material specification sheets available from the material supplier.

Mold Temperatures

When the properly heated plastic is injected into the mold, the cooling process begins. The plastic must be brought back down to a temperature at which it hardens, or solidifies. The faster this is done, the sooner the next cycle can begin. Each type of material has a different rate at which this cooling should occur. Some materials (notably amorphous) can cool very quickly, while others (notably crystalline) require a slower rate of cooling to attain maximum physical values. In either case, the mold itself contains the method of cooling.

Cooling Channels

The standard method of cooling is passing a coolant (usually water) through a series of holes drilled through the mold plates and connected by hoses to form a continuous pathway, as shown in Figure 4-9. The coolant absorbs heat from the mold (which has absorbed heat from the hot plastic) and keeps the mold at a proper temperature to solidify the plastic at the most efficient rate. The hoses are connected to a temperature control unit placed on the floor near the machine. Sometimes, if chilled water is supplied, the hoses are connected to a central chiller that feeds an entire bank of machines and molds.

The cooling process is critical to the molding of quality products and should be metered such that it does not last for too short or too long a time. The overall injection molding cycle consists of many phases, but the major governing phase is the

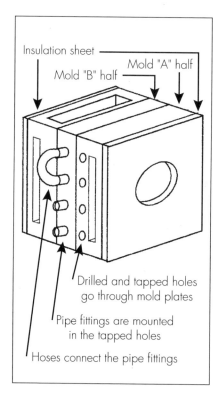

Insulation sheet

Mold "B" half

Mold "A" half

Drilled and tapped holes go through mold plates

Pipe fittings are mounted in the tapped holes

Hoses connect the pipe fittings

Figure 4-9. Cooling channels in a typical mold.

cooling process. Cooling time accounts for approximately 75 to 80 percent of the total cycle.

In an ideal, theoretical world, the mold could be set at a temperature just a few degrees below the melting, or softening, point of the plastic. In the case of polycarbonate, that mold temperature would be approximately 300° F (149° C). However, at that temperature, the overall cooling time would have to be measured in hours, if not days. So, a less than ideal situation is created. In this situation, the mold is held at a temperature that will allow the material to attain satisfactory physical properties within a reasonably short period of time for those properties to be defined. In the case of polycarbonate, that temperature is approximately 180 to 220° F (82 to 104° C). Note that the higher temperature is actually above the boiling point of water. In this case, water could not be used as a coolant and a higher-boiling-point medium must be utilized. Oil is often used as a cooling medium when the mold temperature must be held near the boiling point of water, even as low as 190° F (88° C). When this is necessary, special equipment and hoses must be used to withstand the higher temperatures. By using temperature control systems to keep the mold temperature at a reasonable level, the cooling time can be dropped from a matter of hours to a more acceptable 15 to 30 seconds.

Proper mold temperature control consists of more than just hooking up a mold to a bunch of hoses. To achieve uniform cooling and shrinking of the plastic, the temperature of the mold should be controlled so that any two points on a mold half measure within 10° F (5.6° C) of each other. The way to measure these temperatures is to use a surface pyrometer accurate to within 1° F (0.5° C), and it must have a fast response time. Pick any two points on the cavity image parting line surface of half A or B of the open mold. These points can be within 1 in. (2.5 cm) of each other or at the extreme edges of the cavity image. They should measure within 10° F (5.6° C) of each other. If they do not, the mold is not properly balanced for cooling and hot spots will occur during molding. Hot spots will result in nonuniform cooling and shrinking of the plastic. This will cause stress and warpage, and may also result in parts physically sticking in the mold, because the cooling plastic will tend to stay against the hottest area of the mold.

Cascades (Bubblers)

Sometimes it is difficult to get temperature-control water located where it is needed. An example is in the center of a deep metal core such as those used for making wastebaskets. In those cases, specially designed components can be used. One popular type is called a *cascade* or, more commonly, *bubbler* (Figure 4-10). In a bubbler, the cooling medium (usually water) comes from the main cooling channel, enters at the bottom of the bubbler, flows up through an inner tube, cascades inside the unit, and flows down through an outer tube, exiting back into the main cooling channel.

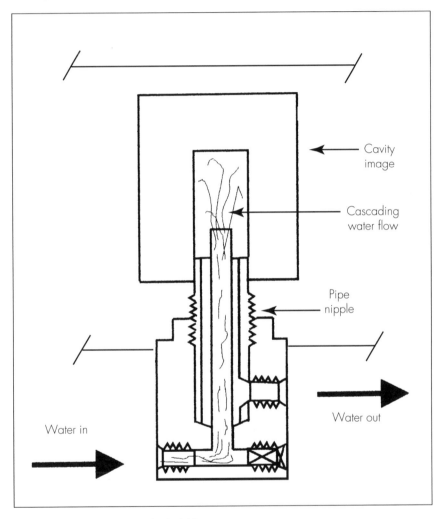

Figure 4-10. Water cascade (bubbler).

Cooling Pins

A similar device is known as a *cooling pin* (sometimes, a *heat pin*). This unit works on the conduction principle and is made from a thermally conductive material such as beryllium copper. In Figure 4-11, the cooling pin is

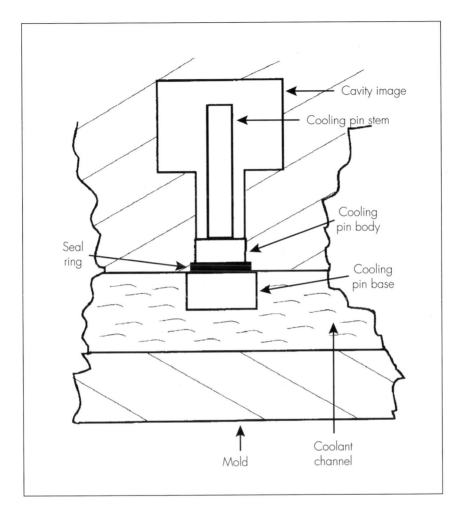

Figure 4-11. Cooling pin.

connected to the metal molding surface and the base of the pin sits in the main cooling channel. Heat is transferred from the plastic to the highly conductive cooling pin. The cooling medium takes heat away from the cooling pin through the base of the pin. In some designs, the pin is hollow and contains a liquid or gas that increases conductivity.

Efficient temperature control also means that both halves of the mold (A and B) should measure within 10° F (5.6° C) of each other. If there must be a difference between the two halves, the B half should be the hotter of the two so that the plastic part will adhere to that half (for ejection purposes).

Insulation Sheets

Consistent and efficient control of the mold temperatures can be aided by mounting insulation sheets on the outside surfaces of the mold. These are 1/4- to 1/2-in. (0.64 to 1.27 cm) thick plates of fiberglass-reinforced polyester (thermoset) that act like asbestos for insulation. Available from mold component suppliers, they are used to keep the surrounding air from influencing mold temperature and help reduce the amount of energy required to maintain proper mold temperatures. They should be mounted on the platen-side surfaces at the very least, but when mounted on all six sides of the mold, they can reduce energy costs by as much as 25 percent.

Cooling Related to Cycle Times

The most important influence on cycle time is the cooling portion of the cycle. The amount of time required for this cooling portion is determined primarily by the average wall thickness of the part and the temperature at which the mold is maintained.

In general, mold temperatures should be set to the values shown in Chapter 3, Table III-2, "Suggested Mold Temperatures for Various Plastics." For materials not shown in that list, the material supplier will recommend the proper starting temperature for the mold.

There are some general guidelines concerning cooling times for any given wall thickness. The first is that if the wall thickness doubles, the cooling time increases by four times. In other words, if a part with a 0.040-in. (0.1-cm) wall thickness has a cooling time of 3 seconds, and the wall thickness is increased to 0.080 in. (0.2 cm), the cooling time goes to 12 seconds. Likewise, if we reduce the wall thickness by half (from 0.040 to 0.020 in. [0.1 to 0.05 cm]), we can reduce the cooling time fourfold, from 3 seconds to less than 1 second. This demonstrates why it is so important to minimize the wall thickness of any part being molded in order to reduce cycle times as well as to cut part weight. Figure 4-12 shows some average cooling times for common plastics.

Cooling Related to Standard Runners

Although cooling times are determined by the average wall thickness of the molded part, a controlling factor is the diameter (thickness) of the runner system and sprue. While these items do not have to be cooled to the same rigidity as the molded part, they do have to become rigid enough for ejection from the mold. Usually, the runner diameter is anywhere from 1/16 to 1/4 in. (0.16 to 0.64 cm) in thickness. The sprue can be up to 1/2 in. (1.27 cm) in diameter at the large end. Both cool and solidify from the outer skin inward and only have to become hard enough to keep from

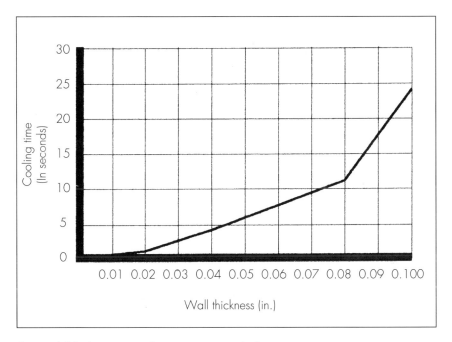

Figure 4-12. Average cooling times versus thickness.

tearing apart on ejection, but the time required for this is usually much longer than for the molded part alone. For instance, a molded part with a wall thickness of 0.060 in. (0.15 cm) would require a cooling time of approximately 7.5 seconds. But if the runner is 0.090 in. (0.23 cm) in diameter, it would require approximately 18 seconds to become fully cooled, and at least 10 seconds to become rigid enough for ejection.

There are some solutions to these problems. The first thing to do is make sure that cavity images are placed as close as possible to the sprue in the mold. That reduces the length that the plastic melt must travel and allows the use of a smaller diameter runner. Next, make sure the hole diameter at the large end of the sprue bushing is no larger than required. This diameter should be one that provides the same cross-sectional area as the total of all the runners coming into it (Figure 4-13).

In Figure 4-13, the runner diameters are both 0.060 in. The cross-sectional area of such a runner equals 0.00283 in.^{2} (0.01826 cm^2). There are two runners, so the areas for both must be added together. This equals 0.00566 in.^{2} (0.03652 cm^2). The next step is to find a diameter that provides an area equivalent to 0.00566 in.^{2}. That would be a diameter of 0.085 in. (0.216 cm). That diameter, then, would be the minimum required for the

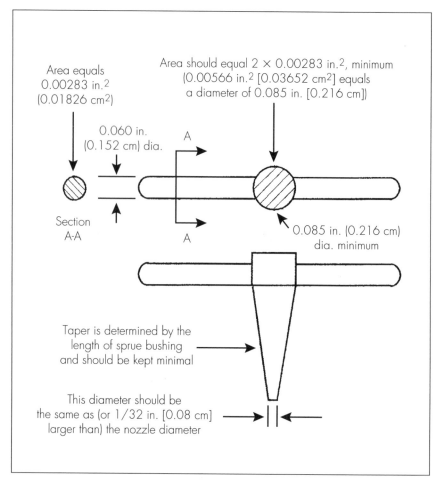

Figure 4-13. Determining sprue diameters.

sprue that feeds those two runners. It could be rounded up (never down) to a common diameter such as 0.09375 in. (0.23812 cm), which is the same as 3/32 in. This ensures that there is always enough material being fed to the runners by the sprue to keep equal pressure on those runners.

The small diameter of the sprue is determined by the opening in the nozzle of the machine, and should be equal to that or slightly (1/32 in. [0.07938 cm]) larger. If it is smaller, the sprue may not align properly with the nozzle and a blow-by condition will occur, which can cause the sprue to stick to the nozzle. Excessive shearing of material in that area could also occur.

Cooling Related to Hot Runners

Another solution to the problem of lengthened cooling times due to runner systems is to eliminate the standard runner system. This can be done by utilizing what is known as a *hot runner*. A hot runner is one that is kept molten during the molding process and thus does not require a cooling time allowance. Figure 4-14 illustrates.

The molten plastic enters the mold through a special sprue bushing, similar to the standard runner mold. But the flow path of the hot runner system is heated (with special heaters) to maintain the molten state of the

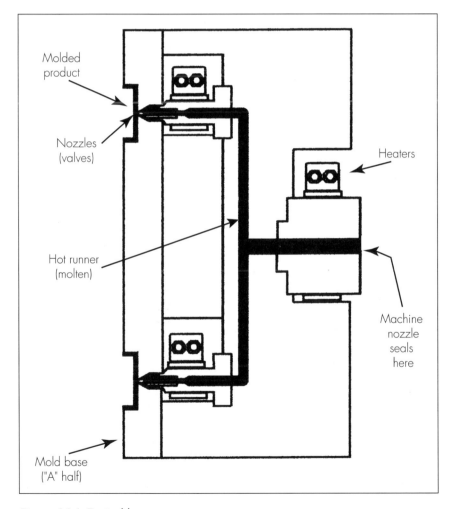

Molded product

Nozzles (valves)

Heaters

Hot runner (molten)

Machine nozzle seals here

Mold base ("A" half)

Figure 4-14. Typical hot runner system.

plastic all the way to the cavity image where the product is molded. Just before it gets to the cavity image, the plastic flows through a special nozzle that allows material to flow until the cavity is filled, then shuts off and keeps the plastic molten and ready for the next cycle.

Several variations of the hot runner concept are available commercially, which enable the molder to eliminate runners and sprues from the total cycle shot. Although the initial cost is high (and can add 40 percent to the cost of a mold), their use will result in shorter cycle times and less leftover material which otherwise must be reground, scrapped, or both.

Machine and Oil Temperatures

The primary area of concern and control of machine temperatures is the temperature of the hydraulic oil in the machine's main system. This is a special oil, designed for use in hydraulic systems, and must be used at a temperature within a range recommended by the oil supplier, commonly between 80 and 120° F (27 and 49° C). If the oil is running lower than 80° F, it becomes sluggish. This causes delays in shifting special valves within the system and can result in machine damage at worst and inconsistent cycles at best. If the oil is running higher than 120° F, it will break down because of thermal degradation. The result can be machine damage caused by loss of additives and lubrication properties, as well as inconsistent, or interrupted, cycles.

Purpose of Heat Exchanger

Heat exchangers are the principal method of maintaining proper oil temperature in a molding machine. They are specialized types of radiators that circulate oil around a series of copper (or other highly conductive metal) tubes that contain circulating water. The water temperature is maintained by special valves that allow more water to flow when the oil temperature begins to rise, and shuts off the water flow if the oil begins to get too cool. The efficiency of these heat exchangers depends on clean, open tubes. Tubes tend to become clogged with scale as a result of minerals in the water, such as iron, limestone, and sulfur. A scale only 1/64 in. (0.039 cm) thick will result in a 40-percent loss of efficiency in the heat exchanger.

Unfortunately, most molding shops do not inspect the heat exchangers on a routine basis and are not aware of the gradual scale buildup until the exchanger cannot keep up with requirements and the machine oil begins running too hot to work properly. The heat exchanger then must be removed and mechanically routed out to remove the scale. This is a costly, time-consuming process, and can be prevented by periodic inspection and cleaning of the heat exchanger tubes. A portable acid-flush unit can be attached to individual heat exchangers on a regular basis (approximately once a month) to keep the tubes clean and scale-free with just a few minutes' operation, without intruding on production schedules. An investment in such a unit will realize immediate payback because of less downtime and more efficient use of water to control oil temperatures. This same unit can also be used to keep mold water lines from scaling, which results in more efficient processing.

Ambient Temperatures

Usually the importance of controlling ambient temperatures is overlooked, or slighted. In fact, the ambient temperature of a molding facility has a great impact on the productivity and efficiency of that facility. It's common knowledge that there are differences between the molding parameter settings of one shift versus those on other shifts. While some may think changes are caused by personnel who think they are capable of running better parts than other shifts, the true cause is usually environmental differences from shift to shift.

Even something as minor as the opening of a loading dock door may have profound impact on the processing parameters of a molding machine. If cooler air is allowed to enter the molding area as a result of that door opening, the temperature control units of the machine may respond by increasing the temperature of the melt. When the door closes, and the room returns to normal temperature, the control units respond again, this time lowering the energy expended. This unexpected cycling of melt temperatures can degrade the plastic material residing in the injection barrel.

Hydraulic oil, mold temperature controllers, heating cylinder controllers, heat exchangers, and a variety of other control units and monitoring systems do respond to conditions immediately surrounding a molding machine. Even fans used for cooling hot areas can cause molds to cool off prematurely. Humid days result in different response conditions to material and mold settings than dry days. And the hot temperatures of summer months can cause machines and controllers to react in a different way than the cold temperatures of winter months.

It is necessary, therefore, to control the molding facility environment if high-quality, low-defect production, coupled with high efficiency and low manufacturing cost, is desired. A positive-pressure air delivery system that changes the facility air approximately four times an hour is suitable. This system must also control humidity and temperature, which should be set and locked and monitored continuously. A relative humidity (RH) of 30 to 50 percent and a temperature of 68 to 79° F (20 to 26° C) are ideal settings.

OPTIMIZING PRESSURE

While temperature parameters may be the most important, pressure parameters are the next most critical. The pressures involved in injection molding are injection pressures and clamp pressures.

Injection Unit

Developing Injection Pressure

In Chapter 3, we saw that the injection process involves three types of pressure: injection pressure, holding pressure, and back pressure.

Injection pressure is defined as the pressure used to perform the initial filling of the mold. Initial filling is usually done at high pressure and speed. The higher the pressure, the lower the melt temperature can be, thus minimizing the cooling time and the overall cycle time. The maximum amount of pressure available depends on the size of the machine and the amount of line pressure it develops. Line pressure is determined by the hydraulic pump system of the specific machine. This is also called *system pressure* and will usually range from 1500 to 3000 psi (10,341 to 20,682 kPa).

Various machine functions use the line pressure to develop pressures for specific applications. Injection pressure, for example, is created by applying line pressure to a hydraulic ram which is located at the back of the injection screw and pushes against that screw to inject plastic into a mold.

Figure 4-15 shows how line pressure is transferred from the pump, to the hydraulic ram, through the screw, and finally to the nozzle and the molten plastic that is ready to be injected into the mold. During this transfer, the pressure is multiplied and increased to approximately 20,000 psi (137,890 kPa). This is accomplished by the mechanical advantage created by the hydraulic oil pushing against the ram, which pushes the screw forward. The pressure is transferred all the way to the front of the screw, at the nozzle. A formula for determining the final available pressure at the nozzle appears in the following paragraphs.

The first step in determining the available injection pressure of any machine is to measure the line pressure developed by the hydraulic pump. This can be found by checking the system pressure gage on the machine, which reads in a range of 0 to 3000 psi (0 to 20,682 kPa). Most machines are manufactured with a standard hydraulic system producing 2000 psi (13,789 kPa) line pressure. This pressure is set at the factory during the manufacture of the machine and should *not* be adjusted. Any adjustments that are necessary should be made on the individual pressure control valves for each parameter, which are fed from the main system pressure.

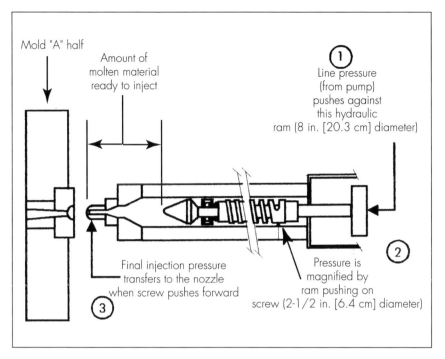

Figure 4-15. Injection pressure on material.

Once the line pressure has been identified, it can be plugged into the following formula for maximum available injection pressure:

$$\text{injection psi} = \frac{\text{pump psi} \times \text{area of ram}}{\text{area of screw}}$$

Assume that the pump pressure is a standard 2000 psi. The area of the ram is found by the formula πr^2, in which π equals 3.1416 and r^2 equals (8 in. ÷ 2) × (8 in. ÷ 2), or 16. Performing the calculations, we find this area to be approximately 50 in.^{2} (322 cm^2). So, the top part of the formula would calculate to be 2000 psi × 50 in.^{2}, which equals a 100,000-lb force (445 kN). Then the area of the 2 1/2-in.- (6.4-cm-) diameter screw is found by the same formula as the area of the ram and calculates to be approximately 5 in.^{2}.

$$\text{injection pressure} = \frac{100{,}000 \text{ pounds}}{\text{area of screw (5 in.}^2)}$$

therefore, 100,000/5 in.^{2} = 20,000 psi

How Much Injection Pressure Is Required?

The maximum available injection pressure for our specific machine then is 20,000 psi. How much pressure is actually required? That depends on what type of material is used.

To begin with, we must first ascertain the flow characteristics of the material being molded. Materials are rated as requiring low pressure (1000 to 5000 psi [6894 to 34,470 kPa]), medium pressure (5000 to 10,000 psi [34,470 to 68,940 kPa]), and high pressure (10,000 psi [68,940 kPa] and above). (A "Melt Flow Index" section in Chapter 6 discusses flow characteristics of plastics.)

The material suppliers rate each of their materials with specific pressure ranges in which to mold. A general-purpose polycarbonate, for instance, should be molded within a range of 8000 to 12,000 psi (55,156 to 82,734 kPa), while a general-purpose nylon 6 should be molded within a range of 1000 to 5000 psi (6894 to 34,470 kPa). So the first thing to do is find what the range is for the specific material being used. If this information is not available from the material supplier directly, a range can be found by consulting any of several buyer guides available through plastics industry publications. Guides provided by trade magazines such as *Modern Plastics* or *Plastics Technology* are valuable tools and contain a variety of good processing information.

Initial Injection Pressure and Time

After the injection pressure range has been determined, a decision must be made as to where to set the pressure initially (within that range) for processing. It is recommended that the pressure be set at the lower end of the range and increased as necessary. This will result in "short" shots to begin with, but will also result in less shock to (as well as wear and tear on) the mold.

In most cases, there will be sensitive components in the mold that can be damaged easily if too much pressure is exerted. Note, however, that there is a definite balance that must be achieved between pressure and temperature. The balance consists of setting the barrel temperatures as low as possible, and increasing injection pressure to cause the material to flow properly. The reason for this is that tests show that plastic materials will exhibit stronger physical characteristics when molded at low temperatures and high pressures. Also, cycle times can be minimized when lower material temperatures are used because the material does not have to cool down as much to solidify again. This is especially true with amorphous materials, less so with crystalline materials.

There are certain problems with molding at the higher end of the pressure range. Excessive pressure may cause higher stresses, especially in the gate area. There is also a tendency of the mold to "lock up," or stick together, because the material has been packed in it so tightly. In addition, if the clamp pressure is borderline to start with, the mold may be forced open, causing flash and short shots. These are all good justifications for starting up by short-shooting the mold and gradually increasing the pressure until the cavities are properly filled. At that point, a 5- to 10-percent increase in pressure can be applied and that final pressure then maintained during the production run.

After the proper amount of injection pressure has been established, the amount of time to inject must be determined. It is good practice to inject material in two stages: an initial injection (or *injection forward*) stage and a holding stage.

Initial injection time should be set just long enough to ensure that enough plastic material is forced into the mold cavities to fill them. Usually this can be performed by moving the material at a rate of 2 to 5 in./s (5 to 13 cm/s) as measured by the indicator flag on the injection barrel. The rate should be set as fast as possible for the product being molded. If booster pressure is available, it should be utilized at this time. As the injection screw approaches the end of the stroke, the speed will slow to approximately 1/16 in./s (0.16 cm/s). This is because the cavities are filled and cannot easily accept more material.

Holding Pressure and Time

If the injecting screw is allowed to return at this point, the still-solidifying plastic will be sucked back out of the cavities by the vacuum created by the returning screw. So, pressure must be held against the plastic until it is solid enough to resist being pulled back into the injection cylinder. Generally speaking, the amount of holding pressure required can be half of that used for initial injection. In other words, if initial injection pressure is 10,000 psi (68,940 kPa), the holding pressure can be set at 5000 psi (34,470 kPa).

The length of time holding pressure should be applied depends on the thickness of the gate the material flowed through, *not* the wall thickness of the molded part. Once the gate freezes off (solidifies), the material in the cavity cannot leak back through the solid gate, and it is safe to release the holding pressure and retract the injection screw. In most cases, the holding pressure time will be on the order of 3 to 6 seconds. Holding pressure time should be reduced steadily until sink marks begin to appear on the surface of the part. This is an indicator that the gate has not solidified and material is escaping from the cavity. At that point, the holding pressure time can be increased in 1-second intervals until the sink marks do not appear after five consecutive cycles, indicating stability.

Cushion (Pad)

It is not possible to know for sure if pressure is being held on the molten plastic unless a cushion, or pad, is utilized. *Cushion* is the term for a slight excess of material that is left in the barrel after the cavities have filled and packed out. This material is that which the holding pressure is applied against. Figure 4-16 shows how a cushion is established. Because of minor inconsistencies in the melt, and shot-to-shot changes in the molding process, it is difficult to maintain a cushion that is set below 1/8 in. (0.318 cm). Therefore, 1/8 in. should be the minimum cushion distance. If the cushion is set for more than 1/4 in. (0.635 cm), the material may begin to lose heat too soon because of the mass of metal surrounding it in the nozzle cap and nozzle. This will cause the material to become semirigid, and it may not inject for the next cycle. Therefore, 1/4 in. should be the maximum cushion distance.

Back Pressure

Back pressure is a force that is used to help homogenize the molten plastic and impart heat to the melt. Homogenizing helps develop consistent density in the melt, and the extra heat helps minimize the amount of heat the barrel heaters must produce. In addition, the heat generated by the back pressure is focused in the center of the melt, which aids in maintaining consistent heat throughout the plastic volume.

Back pressure is created by the turning action of the screw (Figure 4-17). The screw begins turning after the holding pressure control has timed out and the cycle is ready to continue. The next step in the cycle is to prepare the melt for the upcoming cycle's shot. This is done by turning the screw, which augers fresh material forward along the screw flights and into the area in front of the screw tip. As the fresh material begins to build up, it becomes more dense. This causes a slight pressure to develop which begins to push the entire screw backward.

Hydraulic oil that is left in the ram cylinder must be removed to allow the screw to move backward. This is where the back pressure control valve comes into play. It restricts the return flow of oil from the ram cylinder which produces a buildup of pressure in front of the screw tip. That pressure then causes additional mixing and results in an increase in density of the material being prepared. The higher the back pressure setting, the greater the mixing and density. The control valve setting is eventually overcome by the pressure buildup in front of the screw and the screw moves backward.

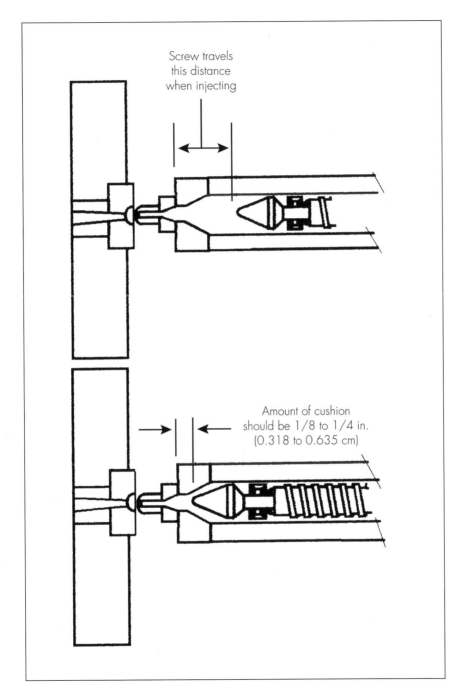

Figure 4-16. The holding pressure cushion.

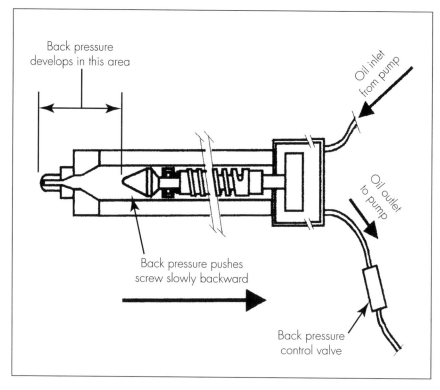

Figure 4-17. Developing back pressure.

The back pressure should be optimized by raising it in small increments of 10 psi (69 kPa), starting with 50 psi (345 kPa). A value less than 50 psi is too difficult to control consistently, and the maximum back pressure value should not exceed 300 to 500 psi (2068 to 3447 kPa). A value greater than that will create thermally degraded material because of excessive shear heat caused by the mixing action of the screw flights. Back pressure should be used sparingly on thermally sensitive materials such as polyvinyl chloride because those materials thermally degrade very easily.

Back pressure settings should begin at 50 psi. The quality of the molded product should be analyzed after a minimum of 10 cycles, and the back pressure should be increased by 10 psi for another 10 cycles. Again, the product should be analyzed and a determination made as to whether additional increases are necessary or whether the back pressure should be reset to 50 psi.

Excessive back pressure will produce unacceptable parts, which are identified by the appearance of such defects as splaying, discoloration (browning), or deteriorated reinforcement. In addition, excessive back pressure may result in a backflow of material over the flights of the screw. This material is molten and may even find its way back to the feed throat. If that happens, bridging will occur.

Decompression

Melt decompression (sometimes referred to as *suck back*) may be necessary if the specific material tends to drool from the nozzle when the mold has opened and back pressure has been applied. The cause of the drooling is a small amount of pressure, built up by back pressure, in the molten plastic waiting in the cylinder (and nozzle) to be injected during the next cycle. This pressure pushes some of the densified melt out of the nozzle as soon as there is nothing to stop it (after the mold opens), if there is no positive shutoff mechanism in the nozzle. Drooling should not be allowed to occur because it will fall across the face of the A half of the mold, cool, begin to solidify, and be crushed when the mold closes. This crushing will damage the faces of the mold plates and result in flashing.

One method of eliminating, or minimizing, drooling is to employ melt decompression. The process involves pulling the screw back slightly (1 or 2 in. [2.5 to 5 cm]) after the screw has returned normally. That is, after injection and while back pressure is applied, the screw returns to a set position to prepare for the next cycle; at that time (before the mold opens), the screw is pulled back slightly farther by 1 or 2 inches. This creates a minor vacuum on the prepared melt and sucks it back into the cylinder and away from the nozzle, thus keeping the drool from exiting the nozzle. This practice does add more air to the flow path and will require additional venting of the mold to eliminate the trapped air during the injection phase.

Clamp Unit

Purpose of Clamp Pressure

The primary reason for using the clamp unit is to keep the mold closed against the pressures generated by the injection unit during the injection phase of the process. A secondary reason is to hold the molten plastic to shape while it cools and solidifies before being ejected from the mold. The longer a product stays under pressure in the mold, the more accurately it will duplicate the finish, shape, and size of the mold.

How Much Clamp Pressure Is Required?

The amount of clamp pressure required for a specific mold is determined by the amount of injection pressure that must be overcome. There are two methods of calculating this, both based on the total amount of *projected area* determined by the shape of the cavity image for the product being molded and of the runner that provides the material to that cavity image (Figure 4-18).

Projected area can be defined as the area of the cavity images and runner layout that is visualized when looking directly at the A or B plate (whichever has the greatest amount). In Figure 4-18, this would equal the two cavity images, as well as the runner system (not including the sprue). We could also visualize the projected area by taking the complete shot that is produced from a single cycle, including parts and runner (do not calculate area of hot runner systems), and measure the area of the shadow

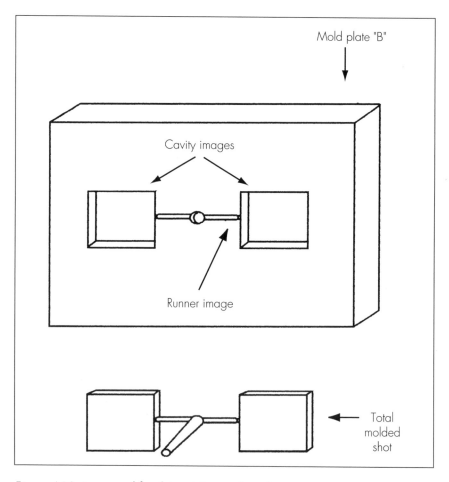

Figure 4-18. Items used for determining projected area.

produced by those items. Projected area is a two-dimensional value and can be thought of as multiplying only length times width. Figure 4-19 shows how the projected area is calculated for the parts from the mold in Figure 4-18.

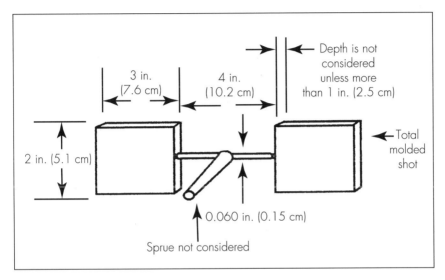

Figure 4-19. Calculating projected area.

In this example, the projected area includes the area of both parts and the runner system. The area of the parts is calculated by multiplying length (3 in. [7.6 cm]) by width (2 in. [5.1 cm]) for each cavity:

$$(2 \text{ in.} \times 3 \text{ in.}) \times 2 \text{ cavities} = 12 \text{ in.}^2 \ (77.42 \text{ cm}^2)$$

Then the area of the runner system is calculated by multiplying its length (4 in. [10.2 cm]) by its width (0.060 in. [0.15 cm]):

$$0.060 \text{ in.} \times 4 \text{ in.} = 0.24 \text{ in.}^2 \ (1.55 \text{ cm}^2)$$

These two results are added together for a total of 12.24 in.² (78.97 cm²) of projected area.

Normally, it is not necessary to deduct holes or other openings that may be present in the molded parts, unless they make up more than 15 percent of the total area. In those cases, that area should be deducted from the total area calculated. Window-frame-shaped parts are good examples of this condition.

Clamp force requirements can now be determined by multiplying the projected area by a factor. This factor ranges between 2 and 8 tons/in.^{2} (27,580 and 110,320 kN/m^2) and represents the amount of force required for each square inch (square meter) of projected area. The lower value of 2 tons/in.^{2} is used for materials that flow easily. The higher value of 8 tons/in.^{2} is used for high-viscosity materials that are difficult to mold and require extremely high injection pressures. A good average value to use is 5 tons/in.^{2} (68,950 kN/m^2). That equates to 10,000 psi (68,940 kPa), which is the average amount of injection pressure used for average-flow materials. In this example, we multiply 5 tons/in.^{2} by 12.24 in.^{2}, which gives a requirement of 61.2 tons (68,950 kN/m^2 × 0.007897 m^2 = 545 kN) of clamp force to keep the mold closed.

Now a machine may be selected. The machine must be capable of producing a minimum of 61.2 tons of clamp force. But, as discussed below, the mold should run in a press that is not capable of creating more than 10 tons/in.^{2} (137,900 kN/m^2) of pressure because it might cause damage to the mold or press. Using 10 tons/in.^{2} on the projected area of our example would require a press capable of creating 122.4 tons (1090 kN) of force.

The second method of determining clamp pressure also requires the calculation of projected area. But, in this case, the projected area is multiplied by the injection pressure that will be used to mold the parts. Injection pressures normally range between 1000 and 20,000 psi (6894 and 137,890 kPa), with the average at approximately 10,000 psi (68,940 kPa). The amount of pressure required depends on the melt flow index of the specific material being molded and the temperature at which the molding is taking place. For our example, we will use 10,000 psi. We multiply that figure by the projected area of 12.24 in.^{2} and get a value of 122,400 total pounds of pressure. Dividing this by 2000 pounds to convert to tons, we calculate a clamp force requirement of 61.2 tons. This is the same value we calculated above using the first method. The rest of the information stays the same. It can be seen by this second method that the clamp pressure requirements can be reduced significantly if the injection pressures can be reduced. That is one reason it is advantageous to mold with reduced injection pressures if at all possible. ("Minimizing Molded-in Stress" later in this chapter addresses this in more detail.)

Now we have established that the mold in our example should be mounted in a press with a rating somewhere between 61.2 and 122.4 tons of clamp force. These figures can be rounded up to the next value of 5, so we can make them 65 and 125 (580 and 1110 kN). We must not go less than our minimum requirement of 65 tons, and we do not want to exceed 125 tons. Any press having a rating somewhere between those two numbers is acceptable for use in this example.

Inadequate clamp force will result in the mold opening during the injection phase of the molding process. This opening will cause flashing and/or short shots. Too little clamp pressure also restricts the degree of adjustment that can be made on pressure parameters.

Excessive clamping will result in mold damage or damage to the press platens. The materials used for making the mold, and the basic structure of the mold design, do not allow the mold to withstand more than 10 tons/in.^{2} clamping pressure without crushing or cracking, and the platens of the machine cannot withstand any more than 10 tons/in.^{2} without collapsing or coining. The higher the tonnage used, the greater the wear on the mold and the press, so it is advantageous to use only that amount of force that will consistently keep the mold closed.

CONTROLLING SHRINKAGE

What is Meant by Shrinkage?

All materials have a specific *shrinkage rate* value assigned to them by the material manufacturer. The term *rate* is actually a misnomer because it implies that the shrinkage occurs as a function of time. Nonetheless, we will use the term because it has become accepted throughout the industry. Shrinkage rate is a value that can be used to predict how much difference there will be between the plastic product when it is first molded and the plastic product after it has cooled (Figure 4-20).

Everything (except water) expands when it is heated and shrinks when it is cooled. Plastic material is no exception. Each plastic material has a distinct value for how much it will shrink after it is heated and then allowed to cool. This value is referred to as the shrinkage rate and is listed as so many *inches per inch* (or *meters per meter*—the values are equal whether in the U.S. Customary System or the International System of Units). That means for each inch of dimension on the plastic product, the material will shrink a certain number of inches. Usually, these materials shrink somewhere between 0.000 in./in. up to approximately 0.050 in./in. Shrinkage can also be thought of in terms of percentage. A part that has a 0.010 in./in. shrinkage will shrink a total of 1 percent. One with a shrinkage rate of 0.020 in./in. will shrink 2 percent. One with a 0.005 rate will shrink 1/2 of 1 percent.

For a shrinkage of 0.010 in./in., we will examine the effect on a part that is 6 in. (15.2 cm) long. Remembering that the shrinkage is for each

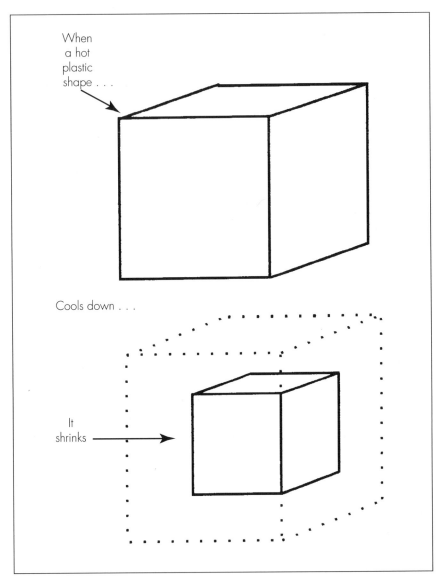

When
a hot
plastic
shape . . .

Cools down . . .

It
shrinks

Figure 4-20. Illustrating shrinkage rate.

inch of product, we would multiply the shrinkage rate by 6 inches. That gives us a total of 0.060 in. (0.152 cm) of shrinkage for that single dimension (0.010 × 6). The mold cavities that will form the finished plastic product must allow for that shrinkage. So the mold maker would make the steel that will form the 6-in. dimension 6.060 in. (15.39 cm). Then, when the material cools, it will shrink to the desired 6-in. dimension.

Note in Figure 4-21 that the mold cavity dimension for the length of the plastic ruler product is 6.060 in. and the width is 0.505 in. (1.28 cm). Assuming that shrinkage is the same in all directions, the plastic that fills that cavity will shrink to 6 × 0.500 in. when it cools, because it has a shrink rate of 0.010 in./in.

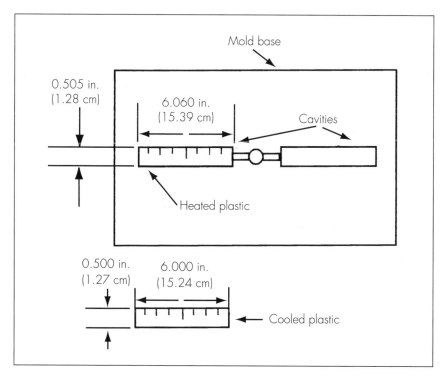

Figure 4-21. How shrinkage affects dimensions.

All plastics are generally categorized as having either low, medium, or high shrinkage. Low shrinkage is within a range of 0.000 to 0.005 in./in. Medium shrinkage is within a range of 0.006 to 0.010. High shrinkage is anything over 0.010.

It is important to understand the difference in shrinkage between amorphous and crystalline materials (Figure 4-22). Amorphous materials tend to have low shrinkage rates and the shrinkage occurs equally in all directions. This is called *isotropic* shrinkage. Crystalline materials tend to have

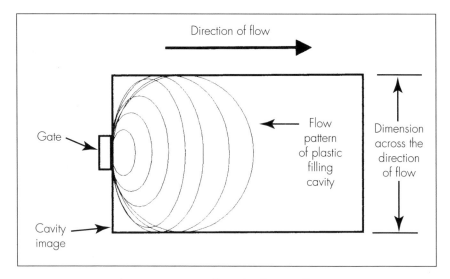

Figure 4-22. Amorphous versus crystalline shrinkage differences.

high shrinkage rates and the shrinkage is greater in the direction of flow than across the direction of flow. This is called *anisotropic shrinkage*. An exception to this anisotropic rule exists in reinforced materials, which shrink less in the direction of flow and more across the direction of flow. This is due to the orientation of the reinforcement fibers.

Because of the inherent differences between amorphous material shrinkage and crystalline material shrinkage, there is a greater range of shrinkage control for amorphous materials. Crystalline materials have a tendency toward higher shrinkage rates in general, but have much less response to processing parameter changes aimed at shrinkage control. The discussion that follows, while general, applies more to amorphous materials than to crystalline.

Effects of Temperature Adjustments

One way of altering the amount of shrinkage for a specific product or material is to adjust the temperature of the plastic while it resides in the barrel. In general, the higher the plastic temperature, the greater the amount of shrinkage. This is because of the activity of the individual plastic molecules; as the temperature rises, these molecules expand more and take up more space. The higher the temperature, the greater the expansion. The reverse of this is also true; the lower the temperature, the lower the degree of expansion, therefore the lower the amount of shrinkage as the plastic cools.

Generally, shrinkage rates can be changed 10 percent by changing barrel temperatures 10 percent. Thus, if a material exhibits a shrinkage rate of 0.005 in./in. at a barrel temperature of 500° F (260° C), it can be lowered to 0.0045 or raised to 0.0055 by altering the barrel temperatures to 450 or 550° F (232 to 288° C), respectively. These are extreme changes and may not be practical for other reasons, but they represent the 10-percent rule of thumb.

Shrinkage can be adjusted by altering temperatures of the mold also. A hot mold will create less shrinkage than a cold mold. This is because the cold mold solidifies the plastic skin sooner than a hot mold, resulting in a shrinking of plastic before full injection pressure is applied. On the other hand, a hot mold allows the molecules to continue to move and be compressed by injection pressure before solidifying. This results in less shrinkage because the molecules are not allowed to move as much after solidifying. A rule of thumb here is that a 10-percent change in mold temperature can result in a 5-percent change in original shrinkage.

Effects of Pressure Adjustments

Injection pressure has a direct effect on shrinkage rates. The higher the injection pressure, the lower the shrinkage rate. This is because the injection pressure packs the plastic molecules together. The higher the pressure, the tighter the molecules are packed. The more they are packed, the less movement they are allowed as they are cooled. This lower movement results in lower shrinkage. The pressure rule of thumb states that a 10-percent change in pressure can cause a 10-percent change in shrinkage rate. Of course, the pressure is applied as long as the material is molten. If the pressure is applied until the plastic has cooled to its point of solidification, the shrinkage will be controlled. If the pressure is relaxed before that point, the shrinkage will increase because the molecules have been allowed to move again.

Postmold Shrinkage

There is a constant battle between maintaining the quality of a molded product and reducing the cost of molding that product. Controlling the shrinkage is only a part of that battle, but it must be understood that the lower the desired amount of shrinkage, the longer the cycle, and the higher the cost. Of course the opposite of this is also true. In fact, under certain molding conditions, once the part is out of the mold, it may continue to cool and shrink for up to 30 days. Admittedly, the first 95 percent of the cooling and shrinking takes place within the first few minutes after removal from the mold, but that last 5 percent can take up to a month. Even if the shrinkage is controlled to achieve that first 95 percent through molding parameter adjustments, the theoretical cycle time could lengthen into 10 minutes for a part that normally runs at a 30-second cycle. One way of minimizing the cycle while controlling the shrinkage is to control the shrinkage after the product is ejected from the mold instead of while it is still in the mold. The cycle time can be reduced, thus the cost of molding can be reduced. This is what postmold cooling and shrinking is all about.

Postmold shrinkage is normally controlled by restraining the molded product in a fixture that holds it in place while it cools. An example is shown in Figure 4-23. Notice that the product is being purposefully bent and bowed in directions opposite the normal shrinking and cooling patterns that develop when a part cools. This is to overcompensate for shrinkage so the part will spring back after cooling to a shape that is desired. This must be done through trial and error by measuring cooled parts to determine how to adjust the fixture to give the desired results.

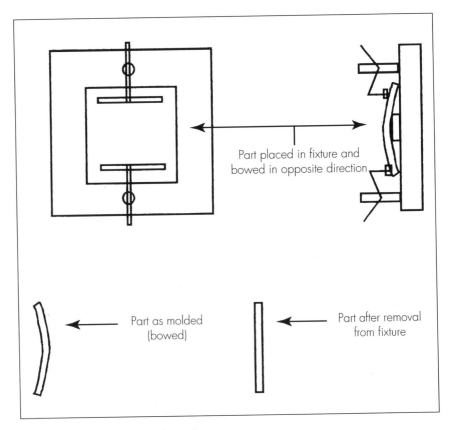

Part placed in fixture and bowed in opposite direction

Part as molded (bowed)

Part after removal from fixture

Figure 4-23. Postmold shrinkage fixture.

When using postmold cooling/shrinkage fixtures, it is necessary to leave the cooling product in the fixture for the equivalent time of approximately six full cycles. Therefore, it is necessary to have a minimum of six fixtures, or stations, in place at all times. Forcing air over the parts helps stabilize them.

Another method of postmold cooling is to simply drop the molded parts in a container of cold water. The temperature of the water must be maintained below room temperature (approximately 60° F [16° C]), but there is no advantage in having it lower than that because once the plastic drops to a temperature below its melting point or glass transition point, it will not continue to shrink. The postmold cooling is done only in an effort to effect that cooling for the center portions of the walls, which take longer to reach the point of solidification than the external skin of the walls.

There is a danger in using any method of postmold shrinkage control because the practice does induce varying degrees of mechanical stress to the molded product. This stress is caused by the forcing of molecules into positions that they are not seeking on their own. When this is done, stress is concentrated on the molecules that are being stretched and compressed, as will be discussed in the next section. This stress is maintained as the part cools and is locked in after the part has fully cooled and shrunk. In this condition, if the part is ever exposed to extreme temperatures or mechanical abuse, the stress is relieved and the product may fracture, crack, or shatter, depending on how much stress was induced during the postmold shrinkage control.

MINIMIZING MOLDED-IN STRESS

Defining Stress

Other than contamination, the single most significant cause of field failure of an injection-molded product is molded-in stress. Stress can be defined as *a resistance to deformation from an applied force* (Figure 4-24). All this means is that if a force is applied to an object, the object resists having its shape changed. The amount of resistance that is present can be identified as stress.

It is possible to understand molded-in stress by visualizing what happens during the injection-molding process. A plastic material is heated to a temperature at which it assumes the consistency of warm honey, and is ready to inject into a mold. During this heating phase, the molecules of the plastic begin to move around. This is what actually causes the material to soften, or melt. Once the material is ready, a plunger device (screw) injects the material by pushing it forward through the machine and into

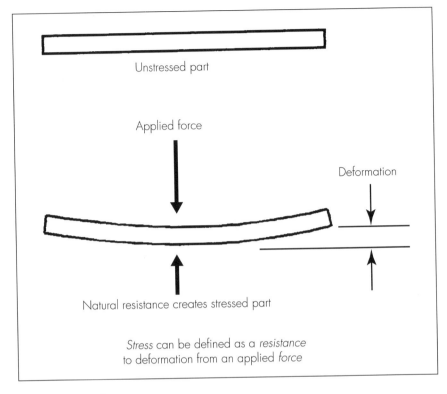

Figure 4-24. Defining stress.

the mold. This pushing action causes the molecules to align, or orient, in a linear fashion. It is similar to pushing a fork through a plate of cooked spaghetti; the pasta strands (molecular chains) start to line up neatly next to each other in the direction that the fork is traveling.

In the molding process, these molecular chains are injected into a mold, where they are then cooled while still being held under high pressure. Because they are kept from relaxing back to their original state, they solidify under stress. It's like stretching a rubber band and then freezing (solidifying) it in that stretched-out position. If the rubber band thaws, it will snap back to its original state. That happens because stress is being relieved. The same thing happens in an injection-molded part. If the part is allowed to relax after solidifying, due to elevated end-use temperatures, or even being knocked sharply against the edge of a desk, the stresses that were molded in can be released, and warpage, cracking, twisting, crazing, or even shattering can occur.

To minimize the possibility of any of these things occurring in the molded part, the amount of stress that is molded in should be kept as low as possible. Although it may not be possible to eliminate it, it is possible to minimize it. This can be done through product design as well as mold design, proper material selection, and proper processing.

Influence of Product Design

Three areas need to be identified as the main causes of stress conditions related to product design: draft angles, sharp corners, and gate location.

Draft Angles

One definition of a draft angle is *the amount of taper required to allow the proper ejection of a molded part from the mold.* It is represented in Figure 4-25.

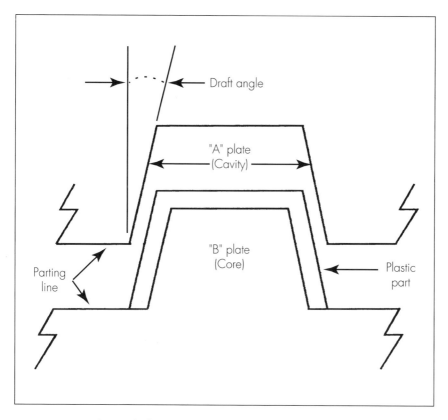

Figure 4-25. Defining draft angle.

How Much Draft Is Necessary?

Ideally, 2° (per side) draft is proper. However, the minimum requirement of 1° may be used, and in some cases, as little as 1/4° may be used. However, the smaller the draft angle, the more difficult it is to get the part out of the mold. It is similar to an ice cube tray. The individual ice cube compartments have tapered walls to effect easy removal of the frozen cubes. If the walls were not tapered, or tapered much less, it would be very difficult to remove the cubes.

It is important to understand that the use of a draft angle alters the dimensions of the part. For illustration, picture a plastic pail. As shown in Figure 4-25, the bottom face of the part is wider than the top face. That is because of the taper of the draft. For every 1° of draft used, the dimension increases by 0.017 in. (0.043 cm) per side, for a part that is up to 1 in. (2.54 cm) deep. For every additional inch (centimeter) of depth, an additional 0.017 in. (0.043 cm) must be added to the dimension. And that is only for one side. Because two sides are affected for each dimension, the increase is doubled. Figure 4-26 shows how the dimensions are affected for a 1° draft allowance.

Note that the part starts out with a width dimension of 2.5 in. (6.35 cm), and that it is 2.0 in. (5.08 cm) in depth. Using a draft angle of only 1° increases the total width requirement of the part to 2.568 in. (6.52 cm) because each degree adds 0.017 in. per side for each inch of depth (0.043 cm per centimeter). Therefore, 0.017 × 2 sides × 2 in. of depth = 0.068 in. (0.043 × 2 sides × 5.08 cm = 0.437 cm) that must be added to the initial dimension of 2.5 in.

What If There Is No Draft?

Figure 4-27 shows the effect of using no draft. Note that a vacuum is created during the molding process. All the air that is trapped in the closed mold during the molding process is displaced by plastic during the injection phase. Whenever air is displaced, a vacuum occurs. When a straight wall exists (no draft taper), the molded part must travel the entire distance of the depth of the molding before the vacuum is released. The amount of pressure needed to push the part out of the vacuum-containing mold is tremendous. While the vacuum tries to keep the part *in* the mold, the ejection force is trying to push the part *out* of the mold. Enormous stresses are set up because of the resistance of the part, in vacuum, opposing the forces being applied by the ejector system. The vacuum resistance may be so great that the plastic is actually punctured by the ejector pins trying to overcome that resistance. The ejector pins may even break off as a result of these forces.

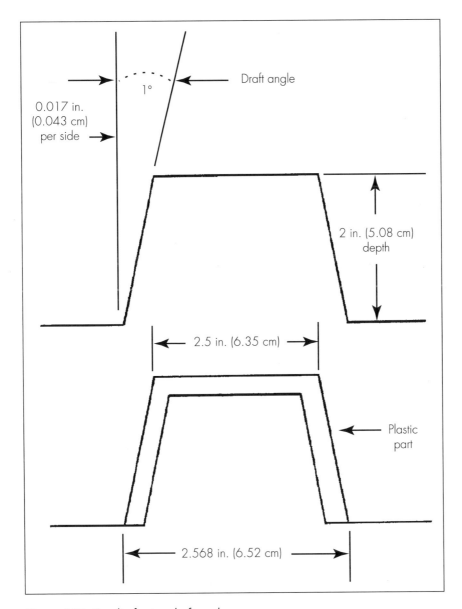

Figure 4-26. Result of using draft angle.

Adding minimum draft to the part greatly minimizes the ejection stresses because the part must travel only a fraction of an inch to release the vacuum. Thus, the use of proper draft will minimize any mechanical stresses that are caused by ejection of the molded part.

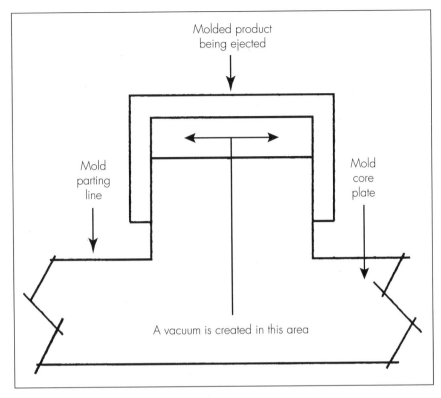

Figure 4-27. Result of not using a draft angle.

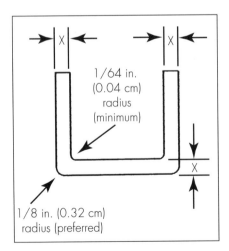

Figure 4-28. Uniform wall thickness.

Uniform Walls

Another way that molded-in stress can be created is by using nonuniform wall thickness. Figure 4-28 shows the cross section of a molded part that does contain uniform walls. Both the sidewalls have the same thickness as the base wall. That makes them uniform. Also, the corners are not squared, but instead are rounded (radiused). This helps make the entire section uniform. Figure 4-29 shows the effect of not having a radius in the internal and external corners. The left

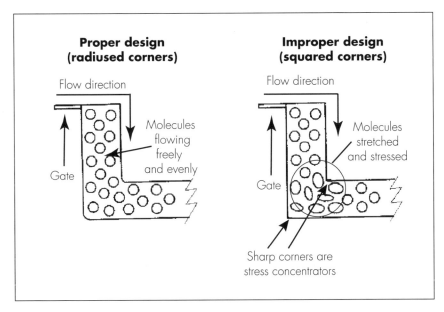

Figure 4-29. Comparing radiused corners to squared.

figure shows a part in which there is a radius in each corner. The molten plastic molecules are injected into the cavity through the gate and flow freely through the wall of the part. They move evenly past the corners and are consistent in shape, and there is no stress applied to them. The right figure shows a part in which the lack of any corner radius has created stress concentration areas in which the molecules are squeezed, compressed, expanded, and sheared as they go around the sharp corner. This stresses the molecules and causes weak areas in the part. With any release of this stress, the part will crack in the corners, or even shatter.

This same condition occurs when there is an abrupt difference in the wall thickness, as shown in Figure 4-30. The distortion of the molecules

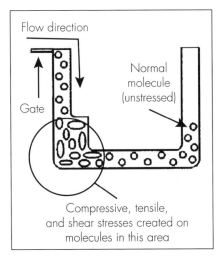

Figure 4-30. Result of abrupt change in wall thickness.

causes mechanical stress to be imparted to the molded product. If a wall thickness *must* change, it should be done with a gentle tapered transition rather than an abrupt change. This is shown in Figure 4-31. Note in Figure 4-32 that the molecules are not stressed and maintain their general shape as they travel through the transition area. The result is a strong part without mechanical stress.

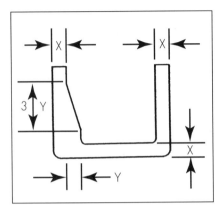

Figure 4-31. Proper transition of wall thickness change.

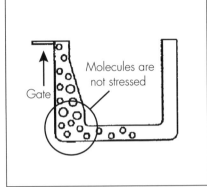

Figure 4-32. Wall thickness change without stress.

THE IMPORTANCE OF DRYING MATERIALS

Moisture is one of the most frequent contributors to defects in molded plastic parts. Any excessive moisture that exists in, or on, plastic pellets during the molding process will turn into steam when the material is exposed to the temperatures present in the heating cylinder of the molding machine. That steam becomes a gas that is trapped in the melt and travels through the flow path into the cavity image and finally becomes molded into the plastic product. The gas appears as silver streaks or droplet-shaped imperfections on the surface and throughout the body of the plastic part. These areas are brittle and are physically weak. Therefore, excessive moisture in plastic materials must be considered a hindrance to proper molding and must always be reduced to acceptable levels prior to processing.

Hygroscopic Materials

There are some materials, called *hygroscopic*, that actually absorb moisture from the surrounding atmosphere. They act like sponges to soak up any available moisture in the immediate area. The most common of these are nylon, acrylonitrile-butadiene-styrene (ABS), and polycarbonate. It is imperative that excessive moisture be driven from these materials just prior to molding, and that they be immediately molded (within 15 minutes) to ensure that they do not absorb fresh moisture before they are exposed to molding temperatures. While most of these materials can be dried successfully in 2 or 3 hours, some of the nylon grades may take as much as 24 hours to obtain the minimum level of moisture required.

Other Materials

It has been stated that not all plastic materials need to be dried before molding. This may be true depending on when the material was first produced, what time of year it was shipped, where it was shipped to and from, where it was stored, for how long it was stored, and what basic type of material it is. But, to make sure there are no moisture problems, it is a good practice to dry *all* materials before molding.

Even materials that are not hygroscopic are susceptible to moisture as a result of condensation. This condensation will be carried into the barrel and turn into steam just like the absorbed moisture in hygroscopic materials. The end result will be the same. So it bears repeating that *all* materials must be properly dried before molding. There is no harm in drying materials, even if they do not need it, as long as drying temperatures are monitored and maintained properly. If they are not, the material granules may become hot enough to stick together (especially amorphous materials) and cause clumping, or even become thermally degraded. Various drying methods, equipment, and processes are covered in Chapter 8.

SUMMARY

Optimizing the parameters for injection molding will ensure quality products, with a minimum of molded-in stress, at lower manufacturing cost.

All parameter adjustments will have an effect (positive or negative) on the physical and aesthetic properties of the molded product. Understanding this relationship allows the molder to manipulate the properties to meet specific requirements established for the product.

Because it takes 6 to 8 hours for parameters to stabilize after a machine is first started, adjustments will be required during that time. One approach to this situation is to have two separate parameter settings—one for initial startup and another for conditions after stabilizing.

Usually, the longest individual amount of time within a cycle is the cooling time. A general rule of thumb is that if the wall thickness doubles, the cooling time increases by four. This demonstrates the need to reduce wall thickness to take advantage of reduced cycle times.

Hot runner systems help in attempts to control and minimize parameter adjustments. They eliminate runners and sprues, which are usually the determining factor in cooling time because they are so much larger than the average wall thickness of the molded product. Hot runner systems also eliminate the buildup of regrind.

Molded-in stress is the single biggest contributor to field defects of molded parts. Usually stress is molded in by excessive injection pressures and thermal degradation of the material, and the stress is sometimes released if the molded part is exposed to high (or low) temperatures or mechanical abuse. This released stress shows up as cracked or fractured parts. Minimizing stress results in fewer defective parts.

Understanding product design requirements assists the molder in adjusting parameters to accommodate design violations. The use of draft angles, uniform wall thickness, and radiused corners will result in lower stress and produce more efficient, less costly products.

QUESTIONS

1. Why is it important to control as many parameters of the molding process as possible?
2. How are part quality requirements normally established?
3. List two property effects that result from:
   (a) Increasing injection pressure
   (b) Decreasing injection pressure
4. Why is it a good idea to have two different setup settings for one production run?
5. In your own words, how would you define *bridging*?
6. At what temperature should the nozzle heater normally be set?
7. What is the main advantage of using insulation jackets on the injection barrel?
8. How is the largest sprue diameter determined?
9. What are the two main advantages to using hot runner systems?
10. What is the definition of *stress* as used in this chapter?
11. What is the recommended minimum amount of draft required for injection molding?
12. Why is draft required?
13. Why should excessive moisture be removed from plastic materials before molding?
14. What is a *hygroscopic* plastic material?

The Role of the Operator　5

FOCUS OF THE OPERATOR'S ROLE

Of all the various components that come together to make up the injection-molding process, the machine operator is by far the most important. All of the equipment, including the machine, the auxiliaries, and the mold, can be fine-tuned and monitored to run flawlessly from cycle to cycle. But the operator is the only component with the capacity to actually think, and therefore can adjust his or her activities whenever needed from cycle to cycle (Figure 5-1). This attribute can be extremely beneficial to an employer because the operator can make on-the-spot observations regarding how well (or poorly) a job is running. The operator is the only part of the production equation that can keep a machine from producing dozens, or even hundreds or thousands, of reject parts.

That being the case, what should a company expect from a molding machine operator? To begin with, the company understands the value of having an operator present during the molding operation. The operator represents the company's interests and is given the responsibility of watching each and every cycle (shot) to make sure that everything is running properly and that acceptable parts are being produced.

Consistency

Of primary importance is consistency. This means that the operator must make sure that every cycle is run exactly the same as every other cycle by operation of the machine gate. The gate is simply a large sliding door that the operator opens at the end of a semiautomatic cycle, and then closes to begin the next cycle. Sometimes the gate opens automatically, but it still must be closed by the operator. When the gate is open, the machine is prevented from closing by electrical and mechanical safety locks. When a cycle is completed and the mold opens, the machine cannot close again until the operator has opened the gate and closed it once more.

The *timing* of opening and closing the gate controls the consistency of the cycle. The operator must make sure that the opening and closing activities take exactly the same amount of time each cycle. A difference of as

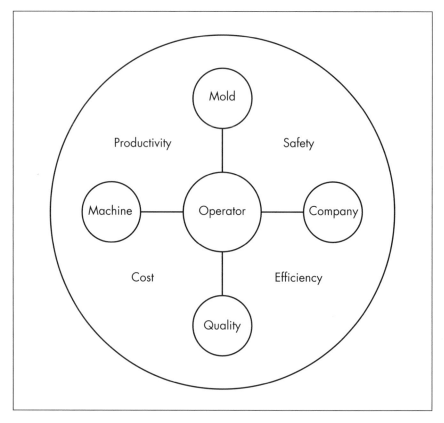

Figure 5-1. Focus of the operator's role.

little as 1 second from cycle to cycle can make a difference of $10,000 or more a year in lost revenue to a company. That's why consistency is such a huge factor in the overall concept of an injection-molding process.

Like any other learned attribute, consistency levels improve with practice. There are some tricks of the trade involved in achieving consistent gate operations, including such practices as counting time while performing the activity, or reciting a poem that has been created to finish at the exact second the activity is completed. These techniques come with experience, and it is expected that a new operator will pick them up from an experienced training operator. That is one of the reasons most companies prefer that a new operator work alongside an experienced operator for some period of time. The amount of training time varies from company to company, but the longer the training, the better the operator will be when the time comes to work alone.

Inspection of Parts

When a molding job is running well, there is little for an operator to do but continue to open and close the gate; but seldom does a job run so perfectly. There is always something for the operator to be doing. For instance, the molded part may not drop off the ejector pins when the mold opens and the operator must reach in and remove the part from the mold before the next cycle starts. Or there may be a small amount of flash present on the molded part and the operator must trim that flash from the part before it is packed away. There may even be some secondary operation required, such as drilling a hole, performing assembly work, or decorating a part.

But, even when the job is running almost perfectly, the operator must visually inspect the parts to make sure there are no defects. Each job should have some detailed inspection information and boundary samples posted at the operator's station so it can easily be seen what is acceptable and what must be rejected. The operator must constantly compare every molded part with that information. It soon becomes second nature for the trained operator to spot defects in a short period of time while inspecting parts. Veteran operators can spot these defects in a fraction of a second, while an untrained person may take several minutes to find the same defect. The deciding factor is level of experience and training.

It is important to understand one major economic fact. Someone pays for every part molded, whether good or bad. The customer pays for good parts, the molding company pays for bad parts. It is imperative that bad parts be discovered quickly so corrective adjustments can be made to the machine settings. It is the responsibility of the operator to make those discoveries and immediately notify the supervisor.

In some companies, the operator is required to make the corrective adjustments. This assumes, of course, that the operator has been properly trained in troubleshooting and molding processing. In such cases, the operator does not inform a supervisor that a change is required, but simply makes the change. But regardless of who makes a change, it must be recorded. This record should include pertinent information such as what change was made; what happened to require the change; the result of the change; and shift information including operator, supervisor, time, and date. All actions should be recorded in a log assigned to that specific mold and filed at the end of the run.

Inspection of the Mold

The mold is an expensive but necessary tool. Its cost can be as high as many tens of thousands of dollars (even hundreds of thousands), so it is important to understand that any damage to a mold can be costly. Some forms of damage may actually ruin an entire mold to the point of requiring total replacement. Molds take from a few weeks to many months to build. So, again, it is important to understand that damage to a mold is costly in terms of time and production that is lost while the mold is out of commission. While the customer is usually the one that pays for building the original mold, the molding company pays for mold repairs.

An operator can help minimize the amount of damage to a mold by simply looking at it every time it opens at the end of a cycle. Some common things to look for are flash, broken pieces of metal, missing components, water leaking from the mold, and part of a plastic part stuck or broken away. There are a variety of other things, but basically an operator should get a good idea of what the mold should look like at the beginning of a job and notify a supervisor immediately if anything at all looks different at any time during the run. If anything does change, the operator should not close the gate, but wait until a supervisor has inspected the mold to determine if it is all right to close it.

Inspection of the Machine

Like any machine, the molding machine consists of many components that must work together. The clamp unit must clamp properly, the injection unit must heat and inject properly, and all the settings must stay within their preset ranges. If anything breaks down, the machine will not operate properly and may produce defective products. Or the machine might cause damage to the mold. Or it may become a safety hazard and cause injury.

The machine operator becomes the company's eyes and ears for detecting any changes in machine operations. The changes can be subtle, such as simple clicking or whining noises. Or they can be very obvious, such as sudden oil leaks or clanging noises. As with the other inspection exercises, an operator must be aware of the normal sounds and appearances of the molding machine operation, and be ready to notify a supervisor immediately of any changes in those normal events. When the machine is operating properly, the sounds it makes eventually become background noise to the operator. They are there, but not really noticed. When a change in those noises occurs, it is like cold water in the face to the operator. The change is very obvious and the operator notices it immediately. At that point, the operator should notify the supervisor for immediate attention.

HOUSEKEEPING

Proper housekeeping at the operator's station is important for at least two reasons. The first has to do with safety. If items are left lying around where

they do not belong, or if tools are not replaced in the proper area, someone may inadvertently be cut with a trimming knife, or trip over a purse
on the floor, or get hurt in other similar ways. Cooling fans may blow
loose items into the face of unsuspecting visitors to the molding machine
station. And uncontained trimmed flash has a way of finding its way into
clothing, causing scratches and possible cuts. In this sense, housekeeping
means keeping the immediate molding machine station orderly and
cleaned up, with all tools properly stored and loose items packed away or
tied down.

The second reason for housekeeping has to do with contamination. This
is the second most common cause of defects in molded parts (second only
to moisture, which is *also* a form of contamination). These defects can be
caused by potato chip salt getting into the plastic material before it reaches
the hopper, or by empty soda cans carelessly deposited in material containers by people who thought they were trash barrels. It can even be
caused by touching freshly molded plastic parts with hands that are oily
or otherwise dirty. Good operator housekeeping practices will help minimize contamination. The wearing of gloves (white linen is preferred) will
help prevent hand oils from affecting parts and will also help to protect
the skin from hot plastic. A small shop vacuum at each press will make it
easier and more efficient to keep the immediate area clean and less dusty.
Good lighting will help illuminate the area, making it easier to notice housekeeping problems.

ATTITUDE

The final area of responsibility for the operator to be concerned with is
attitude. At times, the job of operating a molding machine can seem bor-
ing and tiresome. The job is mostly repetitive, and sometimes the operator feels that no one is paying attention to him or her. In a good, productive
company, this is not the case. Actually, the company wants to listen to the
operator because it is aware that the operator knows better than anyone
else how well or poorly a job is running and how to improve the efficiency
or productivity of a job. The conscientious company will take great pains
to make sure the operator is part of the decision-making process, and will
encourage the operator to make recommendations.

Keeping a bright outlook and cooperative attitude will go a long way
toward ensuring satisfaction in the position of machine operator. Being
quick to notify a supervisor of any changes (as mentioned earlier) and
paying attention to what changes the supervisor makes will give the operator a better understanding of the total injection-molding operation.
Operators must not feel afraid to ask questions. And by all means, they

must realize that they are a vital and important link in the chain that makes up the molding company. No one knows better than the operator the type of quality that goes into the products being molded. The molding company and the final customer both rely heavily on the operator to keep that quality level high, and the production costs as low as possible.

SUMMARY

The most important of the various components that make up the injection-molding process is the operator. The operator can make on-the-spot decisions regarding the quality of a molded part and the efficiency and productivity of a molding process.

A company understands the value of having an operator at the molding machine and therefore requests the operator to be consistent in operations. Consistency is critical to quality and cost.

The operator is also in a position to inspect the molded parts, catching defects before they are sent to the customer. Procedures should also include inspecting the mold for any potential damage because mold repairs can be extremely expensive. The operator is also in a perfect position to inspect the molding machine during the molding process. The operator is the first to notice any unusual noises or actions produced by the machine.

Housekeeping keeps loose material, such as flash, food particles, or dust, from contaminating the raw material or finished part.

The final area of responsibility for the operator is attitude. The company relies heavily on the operator to be conscientious and report regularly on all possible aspects of the job. Input from the operator helps make the production of a molded product successful and profitable.

QUESTIONS

1. What can be considered the most important component of the injection-molding process?
2. What single item controls the consistency of a cycle?
3. Name the three items that should be inspected by the operator.
4. What is the one thing an operator should *not* do if anything seems different?
5. Why is housekeeping by the operator so important?
6. Why is input from the operator so important to the company?

Basics of Materials 6

Depending on which data bank is consulted, there were between 17,000 and 18,000 different plastic materials available to choose from during 1995, with approximately 750 more added each year. Because of the wide range of properties and cost associated with these materials, it is imperative that the material selection process be conducted with appropriate care and attention relative to the finished product's appearance and function.

Material characterization and structure will not be discussed in detail here, but there are important items to be considered regarding the effect of proper material selection on processing. These items, because of their effect on the product during processing, will be addressed briefly.

PLASTIC DEFINED

Plastic can be defined as *any complex, organic, polymerized compound capable of being shaped or formed*. Usually, the terms *plastic* and *polymer* are used interchangeably, although strictly speaking, a polymer is a plastic, but a plastic does not have to be a polymer.

Another distinction that needs explaining is the difference between *thermoplastic* and *thermoset* plastics:

- *Thermoplastics defined.* A thermoplastic is a plastic material that, when heated, undergoes a *physical* change. It can be reheated and reformed over and over again.
- *Thermosets defined.* A thermoset is a plastic material that, when heated, undergoes a *chemical* change and "cures." It cannot be reformed, and reheating only degrades it.

This book, like other volumes in this series, addresses thermoplastic materials in general. They can be compared to water in the sense that they can change from a solid to a liquid, and back again, over and over, without altering their chemical makeup. But, they do need to be placed into one of two categories: amorphous or crystalline.

AMORPHOUS VERSUS CRYSTALLINE

Amorphous Materials

Amorphous (am-OR-fuss) materials are those in which the molecular structure is random and becomes mobile over a wide temperature range. That simply means that these materials do not literally melt, but rather soften, and they begin to soften as soon as heat is applied to them. They simply get softer and softer as more heat is absorbed, until they degrade as a result of absorbing excessive heat. However, it is common and acceptable to refer to amorphous materials as *melting*, so we will do that during our discussions.

Crystalline Materials

Crystalline (CRISS-tull-in) materials, on the other hand, are those in which the molecular structure is well-ordered, and becomes mobile only after being heated to its melting point. That means that these materials do not go through a softening stage but stay rigid until they are heated to the specific point at which they melt. Then they immediately melt. They will degrade if excessive heat is absorbed.

Comparison of Amorphous and Crystalline

Because of their molecular structure, the physical properties of these two types of materials are worlds apart. In fact, they are just about directly opposite each other. Table VI-1 highlights the distinctions.

For each rule there is an exception, and this is true with plastics also. For instance, even though acrylonitrile-butadiene-styrene (ABS) is an amorphous material, it is *not* clear, but translucent. In general, however, the comparisons in Table VI-1 are valid.

Table VI-2 classifies some of the more common plastics.

Table VI-1. Differences in Amorphous and Crystalline Plastics

Amorphous	Crystalline
Clear	Opaque
Low shrinkage	High shrinkage
Softens (no melt)	Melts (no softening)
High impact strength	Low impact strength
Poor chemical resistance	Good chemical resistance
Poor lubricity	Good lubricity

Table VI-2. Plastics Classifications

Amorphous materials	Crystalline materials
ABS	Acetal
Acrylic	Cellulose butyrate
Cellulose propionate	Liquid crystal polymer (LCP)
Polyamide-imide	Nylon
Polyarylate	Polyester (PBT)
Polycarbonate	Polyetheretherketone (PEEK)
Polyetherimide	Polyethylene
Polyethersulfone	Polyethylene terephthalate (PET)
Polyphenylene oxide	Polyphenylene sulfide
Polystyrene	Polypropylene
Polyurethane	
Polyvinyl chloride (PVC)	

POLYMERIZATION

Polymers are formed by combining a series of monomers. Let's look at how this is accomplished.

Monomers

A monomer can be thought of as a single car on a railroad train. The total train can be considered a polymer, and each car can be considered a monomer. A typical monomer is shown in Figure 6-1.

The letters C and H designate that this ethylene monomer is made by combining carbon and hydrogen elements in a specific ratio. We won't worry about how that is done at this point, but note that they are connected to form a single monomer of ethylene. Now look at Figure 6-2. Note that all the ethylene monomers look the same, but they are not connected to each other. Think of them as loose railroad cars, all on different tracks in a train yard, waiting to be linked into a single long train called a polymer.

Polymers

In order to become a polymer of ethylene, the monomers must be connected to each other in a specific way. This is accomplished by exposing them to the polymerization process. After this exposure, they will look like Figure 6-3.

Note that the monomers have all been connected to form a polymer of ethylene. This polymer is called *polyethylene*. So, while a monomer is a single unit of plastic, a polymer is many units of plastic connected in a specific molecular way to create a new chemical compound. This is accomplished through the polymerization process which mixes monomers with a catalyst, and adds pressure and heat to complete the connecting process. Therefore, a definition of *polymerization* might be: *a reaction caused by combining monomers with a catalyst, under pressure, and with heat to form a chain of linked monomers.*

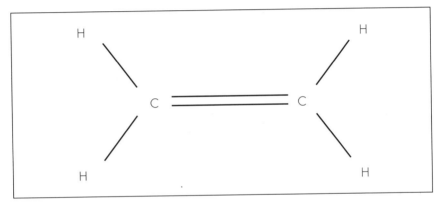

Figure 6-1. An ethylene monomer.

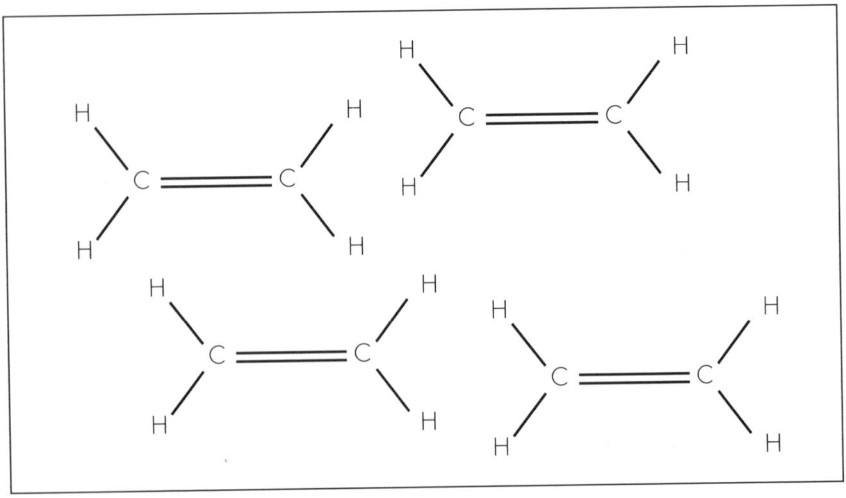

Figure 6-2. Multiple ethylene monomers prior to polymerization.

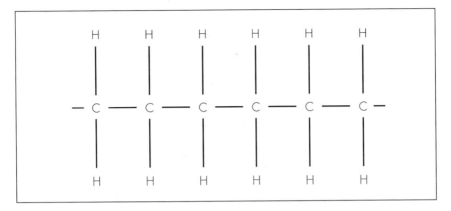

Figure 6-3. Polymerized ethylene monomers.

WHAT HAPPENS DURING THE MOLDING PROCESS?

When considering the action of any plastic material (polymer) while it is going through the injection process, it's enlightening to understand what happens to the molecules. They are bonded during polymerization by forming *chains*. Again, these can be compared to railroad trains where each car represents a molecule and the entire train represents a molecular chain. With amorphous materials, these chains are random, going in all different directions and even crossing over each other. Figure 6-4 shows this relationship. With crystalline materials, these chains are very structured and not random at all. Figure 6-5 shows this relationship.

There are three areas of concern about plastic in the injection process: heat, pressure, and cooling.

Heat

The first thing that happens to the plastic during the injection process is that heat is applied to it. This is done to start the molecules moving (Figure 6-6). As the heat is applied, the molecules begin to move, just as water begins to boil as heat is applied to it. The more heat that is applied, the more the molecules are made to move. We want this motion in the plastic molecules because it means that the plastic is melting (softening). With amorphous materials, we can see the melting take effect as soon as any heat at all is applied, while with crystalline materials, we don't see any motion until the plastic has warmed up to its melting point, at which time it melts all at once. So, heat melts the plastic by setting the molecules in motion.

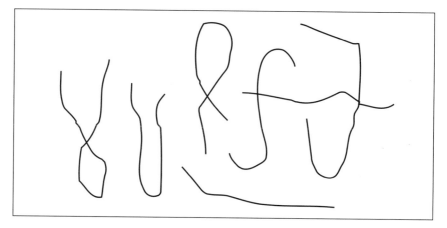

Figure 6-4. Amorphous molecular chains.

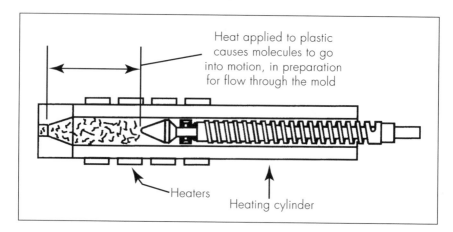

Figure 6-5. Crystalline molecular chains.

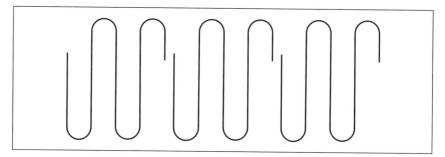

Heat applied to plastic causes molecules to go into motion, in preparation for flow through the mold

Heaters

Heating cylinder

Figure 6-6. Heat sets molecular chains in motion.

Pressure

Once the melting has occurred, we are ready to start the plastic molecules through the flow path and inject them into our mold. We accomplish this by applying pressure to the molten plastic (Figure 6-7).

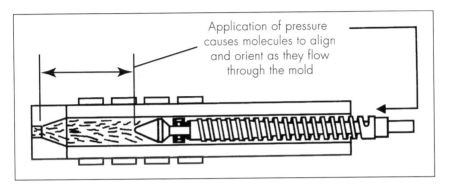

Application of pressure causes molecules to align and orient as they flow through the mold

Figure 6-7. Applying pressure to align molecular chains.

The pressure that is applied actually lines up the molecular chains and orients them so that they can be pushed through the flow path to the mold cavities. As the chains are aligned, they are pushed ahead and go from the injection barrel of the machine, through the nozzle of the machine, and into the sprue bushing of the mold. From there, they go along the runner system, through the gates, and into the cavity images of the mold, where they are packed tightly by the injection pressure. The pressure is held against the molecules in the cavities while the plastic is brought back down in temperature.

Cooling

The cooling process begins as soon as the molten plastic is injected into the mold. The cooling channels of the mold contain a coolant (usually water) that circulates through the mold and takes away the heat brought in by the molten plastic. As the heat is taken away, the plastic begins to cool and eventually reaches a temperature at which the molecular motion stops. This means that the plastic has changed from a fluid state to a solid state and the molded product is rigid enough to be removed from the mold. The plastic cools first at the layer touching the mold surfaces. This forms a skin on the outside of the plastic part where the molecules have stopped moving altogether. Inside, the molecules are still moving, but slowing down. Although the molecules have not completely stopped moving, they have slowed enough to cause the plastic to be rigid on the surface. The internal molecules will not stop moving totally for up to 30 days at room temperature. It is in this way that the cooling action solidifies the plastic by halting (slowing) molecular action.

Summarizing the three phases, we apply:

1. Heat to soften (melt) the polymer by setting the molecules in motion.
2. Pressure to align the molecular chains, causing them to flow.
3. Cooling to solidify the polymer by halting the molecular motion.

COST VERSUS PERFORMANCE

With so many materials to choose from, and the ability of specialty compounders to formulate to order, it is possible to get a material that will exhibit almost any characteristic desired by the designer, the molder, or the end user. There are some caveats to consider, however. As performance increases, so does cost. And, as performance increases, processibility normally decreases. This is shown in Figure 6-8.

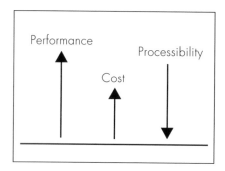

Figure 6-8. Performance versus cost.

Because of this relationship, it is important to analyze the physical, chemical, mechanical, environmental, and thermal requirements of each product design and then select a material that will meet those requirements without incurring prohibitive material costs. Some negotiating may be necessary where tradeoffs in design requirements can be considered in order to reduce overall product costs.

Fillers and Reinforcements

Although the terms *filler* and *reinforcement* are sometimes used interchangeably, there is a subtle difference between the two. Fillers are any additives mixed with a base resin to change the properties of that base resin. Reinforcements are added for increasing only the strength of the base resin. So, a reinforcement could be considered a filler, but fillers are not necessarily reinforcements. It is common to hear the phrase *glass-filled plastic*, indicating that the resin has a glass reinforcement added, but the term should really be *glass-reinforced plastic*.

Fillers are added to a base resin for many reasons. One is to reduce cost. By adding a filler such as talc or clay, resin cost can be reduced because the filler is less expensive than the resin, and the filler adds bulk which reduces the amount of resin required for a specific compound. But, usually, adding fillers *increases* the cost of a plastic compound. Normally, fillers are added to improve a property such as impact resistance, melt flow, flame retardancy, shrinkage rate, or flexibility. In other cases, fillers such as nickel or steel fibers may be used to make the plastic conductive. And sometimes fillers are used simply to add color.

Reinforcements, on the other hand, are additives such as glass fibers, graphite, or mica that are placed in the plastic to impart strength. Tensile strength, compressive strength, flexural strength, and impact strength are all improved by adding reinforcements. In most cases, these are added at a level of 10 to 40 percent, usually by weight but sometimes by volume. In some cases, this may go as high as 70 or 80 percent if the plastic will still be able to flow with such a heavy loading. The addition of reinforcement usually lowers the melt index (flowability) of the plastic. A major concern when using reinforcements is the wear on the mold. Most reinforcements, by nature, are extremely abrasive and can quickly wear out gates, mold texture, runners, or any other area that causes a restriction to the normal flow. For this reason, it is a good idea to use carbide inserts in the gate locations. These will last longer than normal tool steel and, when they do wear out, they can be easily replaced without overhaul of the mold.

As you might surmise, it is quite probable that a plastic will have both filler and reinforcement added to take advantage of the properties of both and enhance the total properties of the plastic.

Melt Flow Index

One of the most important properties of a plastic material is its ability to flow. This is measured by its *melt flow index*, which is a rating of the stiffness of the plastic when heated to a proper molding temperature and injected with a specific amount of pressure. So that we don't have to measure this property under actual molding conditions, a test method has been devised that allows us to test the flowability of a plastic using a small machine on a desk top. This machine, an extrusion plastometer, is commonly called a *melt flow index machine*. It is shown schematically in Figure 6-9. The machine is designed to mimic, or simulate, the action of the plastic under real molding conditions. There is a heated barrel, a nozzle, and a plunger; pressure is mimicked by weight on the end of the plunger.

Every plastic material is categorized in one of many groups for melt flow index measurement. Each category has a set of conditions assigned to it that determines what the barrel temperature should be and what weight should be applied to the plunger.

The test method, ASTM D-1238, consists of placing plastic pellets in the barrel, placing the plunger in the preheated barrel, adding the weight to the plunger, and measuring how much material comes through the orifice during a 10-minute time span. This amount of material is then weighed and the result is listed as grams of plastic extruded in 10 minutes. This is the melt index number. Generally it will fall within a range of 4 to 20 (grams/10 minutes), with an average of 10 to 16.

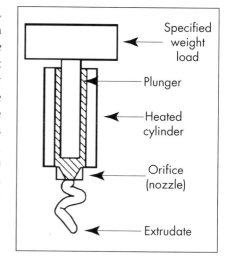

Figure 6-9. Melt flow index machine.

Why Use the Melt Flow Index?

The melt flow index (also called *melt index, flow index,* and *flow rate*) is a valuable tool in that it indicates the flow characteristics of a plastic material *before* that material is used in production. That tells the molder in advance how the material will react while being processed. The flowability of the plastic changes as the melt index changes; so do final properties of the molded product. This is shown in Table VI-3.

Establishing a Proper Melt Index Value

Experimentation will determine what melt index value a specific application requires. A *design of experiments* exercise will help determine the ideal molding conditions for a particular product design and a specific material used. Once this is established, a melt flow test can be performed on the material that was used for the exercise and the result can be listed as the correct melt flow value. This number should be relayed to the material supplier and a stipulation made that no material be shipped outside a range of ±1 of that value. For instance, if the final results show that a melt index of 14 is ideal for a specific application, the material supplier should be advised to ship only material that is between 13 and 15 melt index. The supplier will be happy to comply and will even formally certify the melt index number of every batch of material shipped, if requested.

If a shipment of material does not meet specified melt index values, that does not necessarily mean it cannot be used. What it does mean is that now the molder is at least aware that some major changes must be

Table VI-3. Effect of Flowability on Plastic.

As melt index number decreases:	
Stiffness increases	Tensile strength increases
Yield strength increases	Surface hardness increases
Creep resistance increases	Toughness increases
Softening temperature increases	Stress-crack resistance increases
Chemical resistance increases	Molecular weight increases
But:	
Permeability decreases	Gloss decreases

made to the standard processing parameters for that job in order to utilize the out-of-spec material. If those changes will be detrimental to the quality of the molded products, the molder may reject that particular shipment of material and demand a replacement. If the changes will not be detrimental, the molder may elect to use the material and accept the differences in physical properties and visual appearance of the products.

SUMMARY

Product quality, performance, and cost are dependent on proper material selection. Because there are more than 18,000 plastics to choose from, it is important to understand the molecular structure and behavior of plastics to ensure correct selection for a specific product.

Properties of amorphous and crystalline materials are quite different: this must be understood when selecting and processing plastics.

Polymerization is the process of connecting plastic monomers together, resulting in the production of polymers. Polymers are the materials commonly called *plastics* that are used in the injection-molding process.

The application of three items (heat, pressure, and cooling) to plastic materials is the foundation of the injection-molding process. Heat is used for softening the plastic by setting molecules in motion. Pressure is used to align the softened molecules and push them along the flow path. And cooling is used to solidify the plastic by halting molecular motion.

Any plastic's properties can be altered to satisfy performance requirements. Higher performance requirements result in higher material costs and lower processibility. Therefore, the greater the performance required, the more difficult it is to mold the plastic.

Fillers and reinforcements are used to enhance specific properties or visual features of a plastic product. While all reinforcements can be considered fillers, few fillers can be considered reinforcements. Reinforcements are added for increasing only strength, while fillers alter other properties such as gloss, melt flow rate, shrinkage, and flexibility.

The standard ASTM test procedure D-1238 is referred to as the *melt index test*, which assigns a relative value to any plastic material being used to produce injection-molded products. The test determines the flowability of the plastic and aids the molder in understanding what parameter adjustments must be made, if any, to use a specific batch of material.

QUESTIONS

1. What is the given definition of *plastic*?
2. Explain the difference between *thermoplastics* and *thermosets*.
3. What is meant by an *amorphous* plastic?
4. What is meant by a *crystalline* plastic?
5. Give two examples each of plastics that are categorized as amorphous and crystalline.
6. What two elements are combined to create an ethylene monomer?
7. Define *polymerization*.
8. Why is heat applied to the plastic for injection molding?
9. What is the primary purpose for applying pressure to the plastic?
10. What is the primary purpose for applying cooling to the plastic?
11. As material performance requirements go up, what happens to processibility?
12. What is the primary reason for using a filler in a material?
13. What is the primary reason for using a reinforcement?
14. What is the primary value of using ASTM test D-1238?

Purpose of the Mold 7

DESCRIBING THE MOLD

The injection mold is the heart of the injection-molding process. It is where all the forming action takes place. The molten plastic material, which has the consistency of warm honey, is injected into the mold, under pressure, where it takes the shape of whatever the mold looks like inside (the cavity image). Then the material is cooled until it becomes solid again. When it is ejected, it has the exact image of the shape inside the mold.

In this chapter, we take a look at the injection half of the mold, the A half. Here we see how the plastic material is injected into the mold and how the finished shape is formed. Then we examine the ejector half of the mold, the B half. This is the half from which the finished product is ejected.

The A and B Plates

The mold consists of many different components, but the prime components are the A and B plates, shown in Figure 7-1. Note the two triangle-shaped areas in the B plate. These are called the *cavity images* because they have the shape of the product the mold is going to produce. In this case, that will be two identical plastic triangles with each side 6 in. (15 cm) long, and the parts will be 1/8 in. (0.3 cm) thick. So that is the shape that must be carved (or machined) into the B plate, twice. The A plate will just be flat. However, in some cases, the A plate will also have part (or all) of the image machined into it, if the design of the product so dictates.

The Cavity Image

The cavity image is machined into the plates using standard machine shop equipment such as milling machines, grinders, drills, reamers, and other equipment. Often a machine called an *electrical discharge machine* (EDM) is also used. All of this equipment is capable of producing an image to very exact dimensions, and any image that can be photographed is capable of being copied by machining.

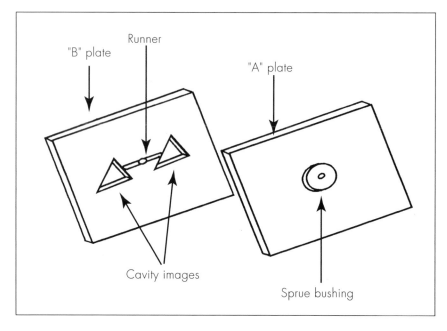

Figure 7-1. A and B plates of a two-cavity mold.

Because of the skills required to operate the machining equipment, and the large investment required to buy it in the first place, the machining of molds is an expensive operation. Finished molds can cost anywhere from a few thousand dollars to a few hundred thousand dollars, depending on size and the complexity of the cavity image. That is why it is so critical that molds be treated properly.

For instance, suppose a molded part does not eject all the way out of the mold and gets stuck. If the mold closes again, that stuck part will likely break some of the ejector pins, core pins, cams, or other fragile components of the mold, costing hundreds or thousands of dollars to repair. Moreover, the mold will have to be removed from the molding machine, losing more money for the molding company because parts are not being produced.

It is imperative, therefore, that the mold be visually inspected every time it opens to make sure the part is properly ejected. At the same time, a quick look will determine if anything is in the mold that should not be there, such as too much flash. If there is a problem of any sort, *the gate should not be closed,* because that will start the cycle over again, and the mold will close on the obstruction.

THE INJECTION HALF OF THE MOLD

The injection half of the mold (Figure 7-2) contains the *sprue bushing* that allows the melted plastic to enter the mold and fill the cavity image, producing the finished plastic part. It is mounted in the center of the locating ring which aligns the mold to the nozzle of the machine's injection unit.

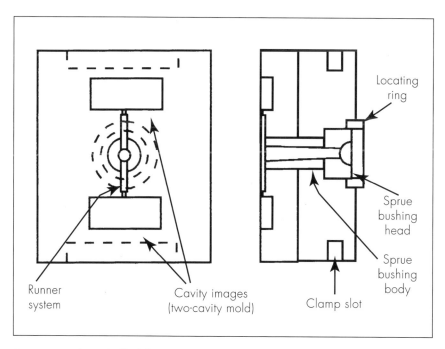

Figure 7-2. The A half of the mold.

The Sprue Bushing

The job of the sprue bushing is to seal tightly against the nozzle of the injection barrel of the molding machine and to allow molten plastic to flow from the barrel into the mold. There is a tapered hole in the middle of the sprue bushing and that is where the plastic flows through the bushing. The hole is tapered to allow the plastic, after it solidifies, to be removed easily to prepare for the next cycle. There are many different basic designs for sprue bushings, the most common being that shown in Figure 7-2. The next most common does not utilize a radiused seat for meeting the machine nozzle, but rather has a flat face. For vertical machines, the sprue bushing is usually split lengthwise and therefore can have a straight internal hole rather than a tapered hole.

Runners

In a conventional mold, the sprue bushing directs molten plastic to the cavity images through channels that are machined into the faces of the A and B plates. These channels allow plastic to run along them, so they are referred to as *runners*. To save material and cycle time, many molds are built with *hot* runner systems (discussed in greater detail in Chapter 4).

Flash

As we discussed in Chapter 4, the clamp unit of the press must exert enough force to hold the mold halves closed during the injection process. Sometimes, if injection values change for some reason, the injection pressure may overcome the clamp pressure. When that happens, the mold will open up a small amount during the injection phase, and a moderate amount of melted plastic will seep out. This seepage is called *flash*. Flash is thin, usually in the range of 0.002 to 0.005 in. (0.005 to 0.013 cm). This is the approximate thickness of a piece of thin writing paper. Although the flash is thin, it can be extremely hard and will cause damage to a closing mold if it flakes off or becomes stuck on the mold surfaces. If a mold is flashing, it should be corrected immediately. However, a small amount of flash may be tolerated as long as it is cleaned off before the mold is closed. This should be allowed only in special cases, and never as a normal operating condition.

As a mold becomes old and worn out, it may also begin to flash slightly. In many cases, this is okay; company managers may have decided that it is more cost-effective to let it flash until they can justify the cost of repairing the mold, as long as no damage is done. In such a case, it may be necessary to remove that flash from the molded product by using a sharp knife-like tool that peels the flash. This is not to be considered normal procedure; the flashing condition must be fixed at some time in the near future or the mold will only get worse and part quality will steadily deteriorate.

THE EJECTOR HALF OF THE MOLD

Ejector Pins

When the mold opens, the finished part is pushed out by a number of ejector pins shaped like nails. They have a head, but instead of having a sharp point at the other end, they are flat. This flat face is what pushes against the molded part. Figure 7-3 shows how they are constructed. The pins have three main areas. The *face* is the part that pushes directly on the plastic product. The *body* is the stem portion of the pin. And the *head* is the part that keeps the pin locked in the mold itself.

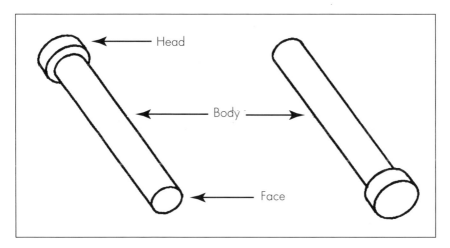

Figure 7-3. Typical ejector pin construction.

Ejector Plates

There are two plates within the mold that lock the heads of the pins and keep them from coming out of the mold (Figure 7-4). These are called the *ejector bar plate* and the *ejector retainer plate*. The retainer plate holds the heads, and the bar plate is bolted against the retainer plate to keep the heads in place. Figure 7-5 shows the ejector plates mounted in the B half of a mold base.

Knockout Rod

Note the *knockout rod* in Figure 7-5. This is attached to the molding machine and enters the mold base through a hole or holes in the ejector housing. There, the rod pushes against the ejector bar plate to advance the ejector pins which push the finished plastic part out of the mold. The knockout rod is also called the *ejector rod*. This design is typical, but there are mechanisms other than the knockout rod that may be used to perform the same function, such as chains, outboard rods, and hydraulic cylinders.

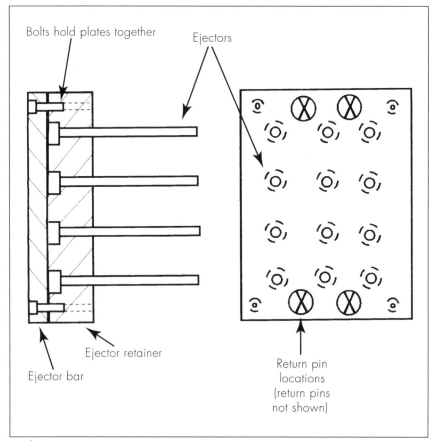

Figure 7-4. Ejector plates.

MOLD DESIGN BASICS

It is important to address a few items of mold design fundamentals concerning gate location, runner design, and venting concepts.

Gate Location

To minimize stress, it is better to locate a gate so that the molten plastic enters the cavity image at the thickest section of the part (Figure 7-6). Although it is ideal to have parts with even wall thicknesses throughout (in which case the gate can be located anywhere), most parts have thin and thick sections.

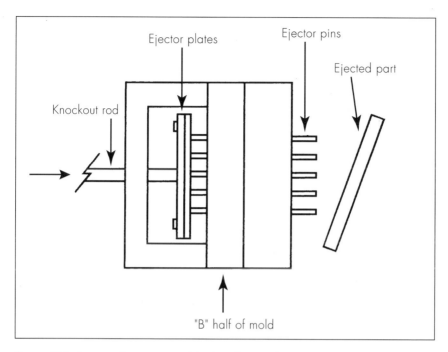

Figure 7-5. Ejector plates actuated by knockout rod.

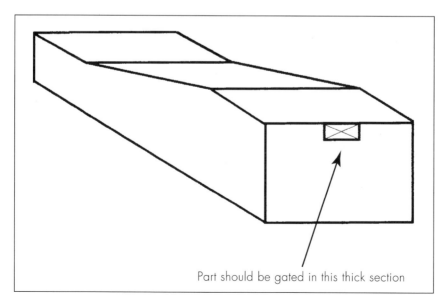

Figure 7-6. Locating gate at thickest section of part.

Based on the discussion in Chapter 4 explaining what happens as molecules flow from a thin section to a thick section, it can be understood that the gate should be located at the thick section to allow the molecules to gently compress, but not decompress. That fluctuation of molecule size is one of the major causes of stress.

Runner Cross Section

A full round runner is ideal. This is because a circular cross section creates equal pressure in all directions on the plastic molecules, while a noncircular section causes unequal pressure. This is demonstrated in Figure 7-7. Using the runner design on the left will minimize the amount of molecular distortion created while the molten plastic is flowing through the runner toward the cavity. Molecular distortion results in stresses in the material; stressed molecules are carried into the cavity, where they solidify in their stressed state.

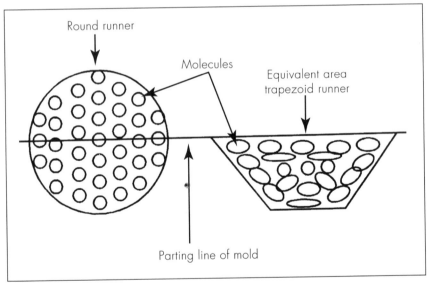

Figure 7-7. Comparison of a round runner cross section with a trapezoidal cross section.

Venting Concepts

There is always a large amount of air that becomes trapped in a mold when the mold is closed in preparation for the injection phase of the molding process (Figure 7-8). This air must be displaced, or removed, so that

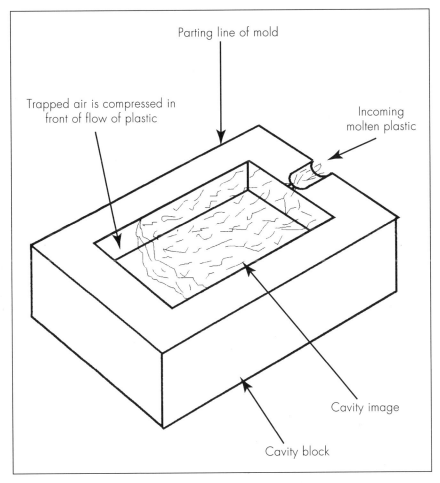

Parting line of mold

Trapped air is compressed in
front of flow of plastic

Incoming
molten plastic

Cavity image

Cavity block

Figure 7-8. Air trapped by incoming flow.

incoming plastic material will be able to fill every available section of the
cavity image and duplicate that image exactly. If the trapped air is not
allowed to escape, it is compressed by the pressure of the incoming material and is squeezed into the corners of the cavity, where it prevents filling
and causes other defects as well. The air can become so compressed that it
ignites and burns the surrounding plastic material.

The most efficient method of allowing the trapped air to escape is to
grind air vents into the parting line of the mold. This cannot be overem-
phasized; suffice it to say that there can never be too much venting. As
long as the vents are of the proper thickness and length, they can be any

width, and they can be any number. A good rule of thumb is to allow at least 30 percent of the parting line perimeter for venting, as shown in Figure 7-9. Measuring the parting line perimeter of this mold shows that there is a total of 10 in. (25.4 cm) of parting line (4 + 4 + 1 + 1). Using 30 percent of that figure as the recommended minimum for venting gives a total of 3 in. (7.6 cm) that should be used for venting. By spacing 1/4-in. (0.64-cm)-wide vents equally along that parting line, there is a total of 12 vents. Common sense dictates that there be vents in corners opposite the gate, with others simply spaced equally from there. The vents could be 1/2 in. (1.3 cm) wide, instead of 1/4 in. In that case, there would be a total of six vents equally spaced. They could be 1 in. (2.54 cm) wide, which means there would be a total of three vents. As long as they total 30 percent minimum of the length of the parting line perimeter, there will be enough vent-

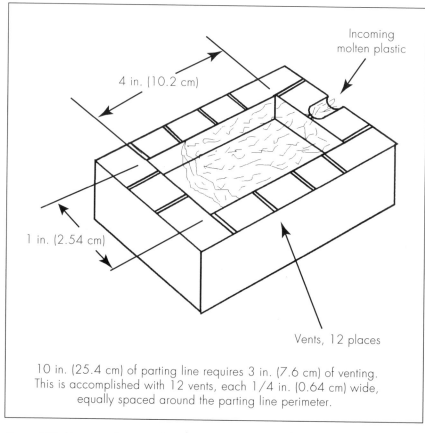

10 in. (25.4 cm) of parting line requires 3 in. (7.6 cm) of venting. This is accomplished with 12 vents, each 1/4 in. (0.64 cm) wide, equally spaced around the parting line perimeter.

Figure 7-9. Venting of parting line.

ing. By using vents that are only 1/8 to 1/4 in. (0.3 to 0.64 cm) wide, more vents can be incorporated and trapped air will have a chance to escape the mold more effectively.

It is also good practice to vent the runner. This eliminates much of the air that is trapped in the runner path from being pushed into the cavity in the first place. A 1/4-in.-wide vent placed every inch along the runner path is adequate. These vents can be staggered so both sides of the runner path are vented, as shown in Figure 7-10.

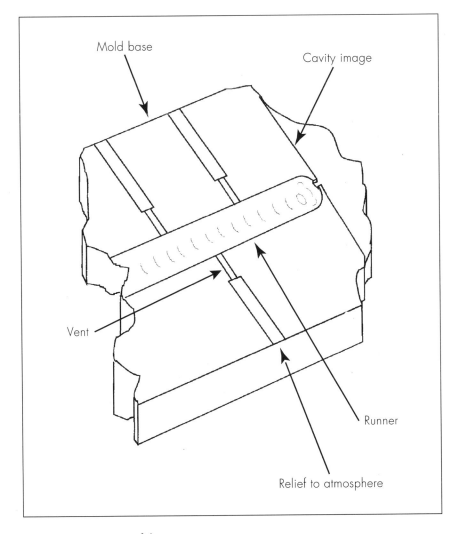

Figure 7-10. Venting of the runner.

SUMMARY

The injection mold is the heart of the injection-molding operation. The plastic molded product is formed within the mold, in cavity images that are machined into the A and B plates.

The injection half of the mold is also called the A half and, in addition to cavity images, contains the locating ring and sprue bushing. These are used to align the mold with the injection barrel of the machine and guide the plastic into the mold.

Hot runner systems are used to eliminate sprues and reduce cycle times.

The clamp half of the mold is also called the B half and, in addition to cavity images, contains the ejection system. Typically, a knockout rod connected to the machine is used to actuate the ejector system of the mold.

Critical areas of concern regarding mold design are gate location, runner design, and venting concepts.

QUESTIONS

1. Why is a mold needed for the injection-molding process?
2. How would you define the purpose of the sprue bushing?
3. Describe *flash* and list two causes of it.
4. What are the two major advantages of using hot runner systems?
5. Name the three parts of an ejector pin.
6. How is the ejector system typically actuated?
7. Where should the gate be located if at all possible?
8. Which shape is best for the cross section of a conventional runner?
9. What causes air to be trapped in a mold?
10. What can be done to a mold to allow trapped air to escape?
11. Why should the runner be vented?

Auxiliary Equipment

8

DRYER UNITS

Dry material is essential to successful injection molding, making dryer units an integral part of injection-molding systems. There are three main types of units for drying material: ovens, hopper dryers, and floor dryers. Vented barrels are built into specific brands of molding machines and are not considered auxiliary equipment; however, it should be noted that vented barrels cannot remove all of the moisture necessary to ensure proper molding, and they should be used only in conjunction with one of the other three drying methods.

Regardless of the type of dryer, once material has become dry enough to mold, it will stay dry for only about 2 to 3 hours. Therefore, it is not practical to dry material much more than 2 hours ahead of the time it will be used. In fact, most machine hoppers are designed to hold approximately 2 hours' worth of material on the average, for that very reason. The use of hopper dryers will extend that time because the hopper dryer unit continuously dries the material within it, and has an extended-size hopper.

Hopper Dryers

Hopper dryers are the most common units for drying and maintaining dryness of plastic materials prior to molding. They work on the principle of circulating warm, dry air through a mass of plastic pellets (Figure 8-1). Dry air absorbs moisture from the pellets and is taken away, back to the dryer unit where the moisture is deposited into a bed of granular absorbent material called a *desiccant*. Common desiccants are calcium chloride and silica gel. After a few hours use (6 to 8), the desiccant, which has become saturated with moisture, is removed and regenerated by placing it in a high-temperature oven that drives off the moisture and freshens the desiccant for future use. This recycling process can be built right into the main system to perform automatically. Normally, a bank of desiccant containers is employed so one can be regenerating while another is in use.

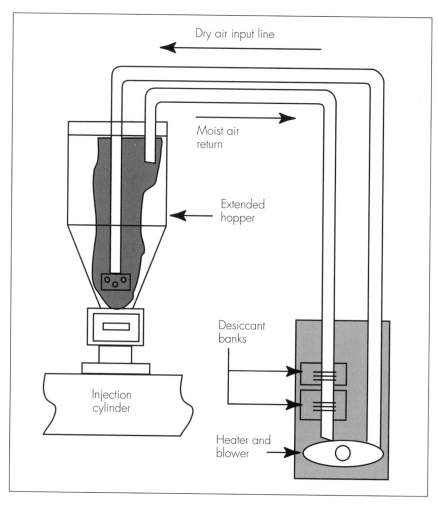

Figure 8-1. Typical hopper dryer operation.

Floor (Central) Dryers

Using hopper dryers on every press can become expensive for a molding operation having a large number of molding machines. An alternative is to use a central drying system (Figure 8-2), commonly referred to as a *floor dryer*. This is similar in operation to the hopper dryer except that one central unit feeds a number of machines. Commonly, four to six machines can be served by a single central dryer. There are some very large units for use with a large number of presses, but their cost is usually prohibitive for most operations.

Figure 8-2. Floor dryer unit. (Courtesy AEC)

Floor drying units operate by one of two basic methods. The first pre-heats and dries the material in a central spot and then feeds that warm, dry material, as needed, directly to the machine using it. In this way, small storage hoppers can be utilized for each machine, holding approximately 30 to 60 minutes' worth of material, which minimizes the tendency of the material in the hopper to pick up moisture. Such a system is practical only if the same material type and color is being molded on all the machines fed by the dryer unit.

The second method is to pump dry air to each machine from a central supply. After drying, extended hoppers store 2 to 4 hours' worth of material. With this system, any number of materials can be dried at the same time. In this case, the dry air is not heated, and a preheater is required at the hopper of the machine.

Oven (Drawer or Tray) Dryers

Oven dryers are the original drying units, dating back to the 1920s. These units consist of a series of trays (drawers) mounted on a rack within a closed chamber (Figure 8-3). The chamber forces hot, dry air over the trays. The trays are filled with plastic pellets to be dried and usually hold from 25 to 50 lb (11 to 23 kg) each. The pellets are poured into each tray to a depth of 1 1/2 to 2 in. (3.8 to 5.1 cm). As the dry air flows over the trays, it picks up moisture and transfers it to a desiccant bed. The dry air is returned to the chamber for another pass. As with the other drying units, oven dryers require the desiccant to be regenerated on a timely basis, usually every 8 hours.

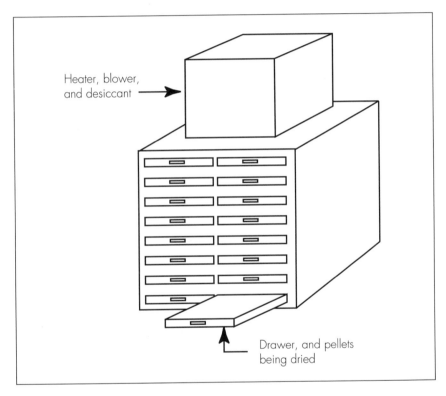

Heater, blower, and desiccant →

Drawer, and pellets being dried

Figure 8-3. Typical oven-type dryer.

How Dry is "Dry"?

Each plastic material requires a specific amount of drying, at a specific temperature, and for a specific amount of time in order to reduce moisture content to an acceptable level. For instance, acrylonitrile-butadiene-styrene (ABS) material must be molded with a maximum moisture content of 0.10 to 0.15 percent, by weight. This can be achieved by drying at 200° F

(93° C) for a period of 2 hours, in a standard dehumidifying dryer. Some nylons, however, may require up to 24 hours' drying at 190° F (88° C) to achieve acceptable moisture levels. Temperatures and drying times may be obtained from the material supplier. But how can you tell if the right moisture level is achieved?

Dew-point Measurement

Dew point is a temperature at which the plastic material is considered dry when a dehumidifier is used. A special meter, available from supply houses, indicates the dew-point temperature. For most plastics, the dew point must fall in the range of −45 to +10° F (−43 to −12° C). The actual requirement for a specific material may be obtained from the material supplier. The dew-point meter can then be used as a monitor to ensure the plastic has been thoroughly dried before it is processed.

Inexpensive Moisture Testing (TVI Testing)

An inexpensive, yet accurate, method of testing materials to be sure they are dry enough to be molded was developed by the GE Plastics Section at Pittsfield, Massachusetts. Called the Tomasetti volatile indicator (TVI) after the GE applications engineer who developed the technique, the simple method requires very little equipment and is performed in six easy steps.

The equipment consists of an electric hot plate capable of maintaining 525° F (274° C), two standard glass microscope slides, tweezers, and some wooden tongue depressors. The procedure is as follows:

1. Place the two glass slides on the surface of the hot plate (which has been preheated to 525° F) for 2 minutes.
2. After 2 minutes, use the tweezers to place two or three plastic pellets on the top of one of the glass slides (Figure 8-4).
3. Now, using the tweezers, place the other glass slide on the plastic pellets, making a sandwich of two glass slides with the pellets between them.
4. Using the long edge of a tongue depressor, press the slide sandwich together until each pellet flattens out to about a 1/2-in. (1.2-cm) diameter.
5. Using the tweezers, remove the glass slide sandwich from the hot plate and allow it to cool approximately 5 minutes.

Figure 8-4. TVI test slide. (Courtesy GE)

6. Calculate the approximate ratio of the total area of any bubbles formed to the total area of the plastic pellets (Figure 8-5). This ratio can be interpreted as the percentage of moisture present in the original pellets.

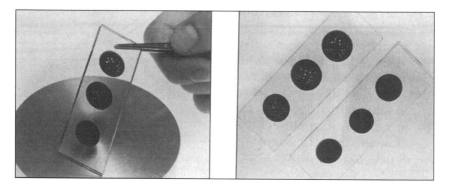

Figure 8-5. Examples of TVI results. (Courtesy GE)

This test gives a close approximation of moisture content, which can be compared to the material supplier's figures to determine if the material is dry enough to mold. The fewer the bubbles, the dryer the material.

It should be noted, however, that this test is not accurate for glass-reinforced plastics, but it can still be used as a reference test for them. In addition, if only one or two bubbles is found, it may indicate trapped air instead of moisture, so the material can be considered dry.

LOADERS

Every injection-molding machine depends on a consistent supply of fresh plastic material to be used during the molding process. This material must be moved from a storage container, after being dried, to the machine hopper so the machine can process it. Handling the raw plastic material, usually in pellet form, can become time-consuming and labor-intensive, especially as the number of molding machines increases. So, this material handling phase of the molding process tends to be performed by machinery whenever possible.

The choice of material handling equipment depends on four factors:
1. The type of material (usually pellets, but sometimes powder).
2. The amount of material required for consistent processing.
3. The vertical and horizontal distances being covered.
4. Special functions required beyond movement (such as mixing color).

Mechanical Loaders

The most common mechanical loader is the auger type, which employs a long-pitch, spring-shaped auger rotating within a tube. The discharge end of this tube is attached to the top of the molding machine hopper, and the pickup end is placed inside the material shipping (or storage) container, usually placed next to the molding machine. When the system is activated, the rotating auger picks up material from the container and carries it through the tube and into the machine hopper. Activation can be initiated by a timer or level sensor, or manually. While this is an inexpensive transfer system, it is very difficult to clean and should be considered only when one specific material will be used for an extended period of time.

Another method of mechanical loading (although not commonly utilized) is the elevator loader. This system consists of an elevating platform placed next to the molding machine in such a way as to raise a container of material up to the level of the hopper. At that point, the elevator table tilts and pours material from the container into the hopper. Then it lowers the container back to floor level. To be effective, this system requires an extended size molding machine hopper and a large amount of floor space, so its use is fairly limited.

Vacuum Loaders

By far the most popular loading system for transferring raw plastic to the molding machine hopper is the vacuum loader (Figure 8-6). The most common version of these consists of an integral motor (or pump) unit mounted directly on top of the molding machine hopper. Tubes are connected to the unit and placed within the material container up to 50 ft (15 m) away from the machine. When activated, the unit creates a vacuum that sucks material from the container and pulls it to the top of the hopper, where the unit dumps the fresh material directly into the hopper. These systems are suitable for applications requiring material loading rates of up to 2000 lb (907 kg) per hour.

Positive-pressure Loaders

When many machines are using the same material, or very large volumes of material are being used by various machines, it is possible to utilize large positive-pressure loading systems that are centrally located. They are connected to each molding machine by way of tubes and can transfer material over great distances (up to 500 ft [152 m] or more) so the relatively high cost can be spread over a large number of machines. These systems can also be designed so that each molding machine can be fed a

Figure 8-6. Vacuum loader. (Courtesy Polymer Machinery)

different material, or color of material, without cross-contamination. Each system can deliver approximately 15,000 lb (6804 kg) of material an hour, and are typically called on for unloading bulk railcars for silo storage.

BLENDERS

Blending may be defined as combining two or more types of materials to give a uniform mixture. For injection molders, blending may be required to achieve proper color combinations or to combine regrind with virgin pellets. It can also be used for adding ingredients to improve flow, reduce sticking, increase flammability resistance, or enhance the base material in a variety of other ways.

Manual blending can be performed simply by stirring a predetermined amount of the desired ingredients into a measured batch of raw plastic pellets. It can be as simple as rolling a barrel of the combined materials across the floor for 30 minutes. But, although simple, the labor costs and potential for inconsistent blending of manual methods have led to the popular installation by injector molders of automated, integral blending systems.

Blending units are actually metering devices that allow a specified amount of ingredient (usually pellets or powders, but sometimes liquid) to be combined with a specific batch of plastic material just before it is dumped into the molding machine hopper. The most common method of automated blending is performed by a unit mounted atop the hopper loading unit on the molding machine (Figure 8-7). Some loaders have built-in blending units.

Figure 8-7. Machine-mounted blender. (Courtesy AEC)

When a product generates a great deal of regrind, it is advisable to utilize a blender to mix that regrind with incoming virgin material in an effort to use up the regrind as it is generated. In these cases, it is best practice to connect a blender pickup hose to the storage compartment of the regrind granulator so regrind can be used immediately after initial molding, before it has a chance to absorb moisture from the environment.

GRANULATORS

Granulators consist of a rotary shearing cutter, driven by an electric motor that moves across a heavy metal screen (Figure 8-8). The plastic to be granulated is placed in a hopper that feeds the plastic to a cutting chamber (Figure 8-9). There, the rotating cutters chop the plastic into small particles that are continuously cut until they fall through the predefined openings in the screen.

The shearing action is provided as a result of the cutter blade rotating past the stationary shear anvils. Large chunks of regrind material are sheared from the plastic product and these are further sheared until they are small enough to fall through the perforated screen. Because these particles are irregularly shaped, screens have been developed in various sizes to accommodate a variety of particle sizes ranging from approximately 1/8 to 5/16 in. (0.3 to 0.8 cm).

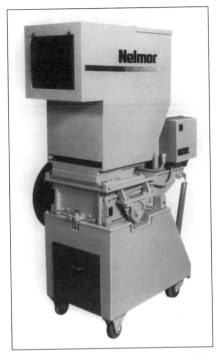

Figure 8-8. Granulator. (Courtesy AEC)

Often, granulators are used to feed regrind directly back to the molding machine hopper by way of a hose (tube) connected to an opening in the bottom of the granulator's storage container. The regrind is then metered into the virgin material at the hopper by a blender unit, such as described in the previous section.

Sometimes regrind operations are performed at a location away from the press. This is generally done at a central location (especially if the same material is used for a number of molding machines) so that a large granulator can be used to grind a large number of parts. The regrind material can be stored for future use, or cycled back to specific machines for immediate integration with virgin material.

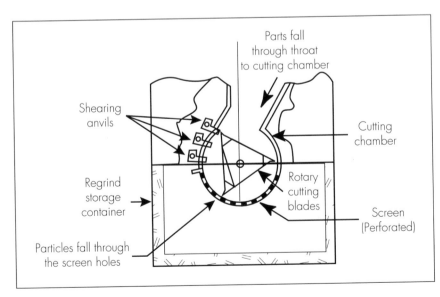

Figure 8-9. Granulator cutting chamber.

MOLD TEMPERATURE CONTROLLERS

For efficient, productive molding, the mold temperature must be controlled as closely as possible. This can be accomplished by a mold temperature control unit (Figure 8-10). A variety of sizes are available, with some units mounted at waist height and some at floor level, but in most cases, the units are mounted on casters and can be moved easily from area to area. The unit is electrically operated and must be plugged into a proper outlet. It must also be connected to a water supply or other coolant source.

As shown in Figure 8-11, a supply hose is connected from the unit to one half of the mold, with a return line from the mold half to the unit. Though not as efficient, some molders use a single unit to control both halves (A and B) of the mold. This requires that one half of the mold receive supply water from the other half, which in turn means that the incoming water is usually hotter for the second half. That makes it extremely difficult to maintain a temperature difference within 10° F (5.6° C) between the two mold halves, as required for efficient production. It has proven to be much more effective to utilize two control units per mold, one for each half.

The operation of the unit is simple. A liquid, usually water, is circulated through the mold by the control unit. This circulation is accomplished by sending the water through a hose connected to one half of the mold. The return hose circulates the water back to the control unit. The control unit senses the temperature of the water at that time and compares it to the preset temperature required for that mold. If the water is too hot, the unit drains some of it and replaces it with fresh, cold water until the temperature setting is matched. If the water is too cold, an internal heater in the unit warms it to the tempera-

Figure 8-10. Mold temperature control unit. (Courtesy AEC)

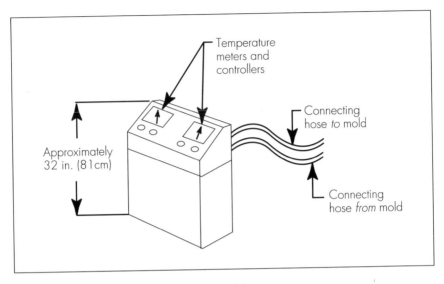

Figure 8-11. Connecting control unit to mold.

ture setting before it goes back to the mold. Units without an internal heater simply stop circulating the water until it has absorbed enough heat from the mold itself.

The temperature-controlling process is continuous, requiring the control unit to operate in a fluctuating manner, alternately heating and cooling the water to maintain the proper temperature.

The most common coolant liquid is water, but when temperatures above the boiling point of water are needed, a liquid such as mineral (or silicone) oil is used because it has a much higher boiling point.

ROBOTS

Use of robots in the injection-molding industry has boomed in recent years. The main reason for this is not so much to reduce labor costs as to achieve consistency in machine cycles. Robots are unique in that they do not require breaks or lunch periods, and they seldom miss work due to illness. Also, it makes more sense to use a robot for tedious, repetitive, physical labor and employ humans for jobs requiring thought, analysis, and decisions. In the robot world, there are two basic types: rigid and flexible.

Rigid robots are those designed and built to perform tasks associated with only one specific application. These robots actually look like sophisticated machines; they might drill holes, machine shapes, or perform similar work. They are usually bolted in place and are not designed to be moved or adjusted.

Flexible robots are those designed to be moved and adjusted to perform various activities that may change from product to product. Examples of these are robots that remove finished parts from molding machines. They can be moved from machine to machine and are fitted with quick-connect fittings at the end of their "arms," which can receive any number of tools or grasping attachments. These robots are also called *pick-and-place* units because their primary function is to grasp some item, move it to another location, and place it on a table, in a fixture, on a conveyer, or in a box. They can also be used to perform rudimentary operations such as snapping two parts together.

SUMMARY

Dry material is essential to successful injection molding. There are many styles of drying units, including hopper dryers, oven dryers, and floor dryers.

Although dew-point measurement is a good way to test for proper dryness, an inexpensive test called *TVI moisture testing* is extremely accurate.

Loaders are machines used for placing raw plastic in machine hoppers. They can be mechanical, vacuum, or positive-pressure pneumatic devices.

Blenders are utilized to mechanically mix additives with raw plastic pellets. Mixing can be done prior to loading the material in the hopper or at the hopper itself.

Granulators are machines that grind up scrap and excess molded material (such as runners) to create regrind.

To maintain the proper temperature for the mold, a mold temperature control unit is used. This unit contains a coolant (usually water, sometimes oil) that is circulated through the mold; the temperature of the coolant determines the temperature of the mold.

Robots are being used in the injection-molding industry to perform tedious, repetitive activities such as machine operation.

QUESTIONS

1. What are the three main types of units used for removing moisture from material?
2. Besides dew-point measurement, what test is used for determining moisture content?

3. What are the three material-moving methods employed by loading machines?
4. How far (in feet [meters]) can a vacuum loader transport plastic pellets?
5. What is the primary purpose of a blender?
6. What is a granulator used for?
7. How many temperature controllers should be connected to a single mold?
8. Name the two common media used in mold temperature controllers.
9. What are the two basic types of robots?
10. Into what category does the pick-and-place robot used in injection molding fall?

Secondary Operations　9

DEFINING SECONDARY OPERATIONS

For our purposes, a secondary operation is defined as *any operation that is performed on a product after it has been molded*. Such operations normally include, but are not limited to, assembly, machining, and finishing (including decorative finishing). In this chapter, we look at the fundamentals of these secondary operations.

It should be noted that secondary operations usually mean increased product cost. Most secondary operations are performed outside the normal molding cycle time and usually require additional personnel. In some cases, the secondary operations are performed right at the molding machine by the machine operator if time allows it within the normal molding cycle time.

Secondary operations can be eliminated by proper part design and proper mold design. However, this too comes at a cost: *ALL secondary operations can be eliminated through part design and mold design, IF cost and time are not a consideration.*

For instance, a part can be decorated within the mold by using existing technology that actually deposits the required finish on the mold surface, and this is transferred to the product during the molding cycle. Another example would be assembly. In this case, two parts are molded side by side and snapped together while the mold is ejecting them. Or, one part can be molded, then insert-molded into the second part. In such a case, the cost of designing and building the secondary capabilities into the mold are expensive and can add 200 to 1000 percent to the normal mold costs. It can also add the same percentage to the delivery time for completing the mold build. These factors must be carefully considered when determining whether to build the secondary operations into the mold or perform them outside the mold.

Another consideration is the use of robotic systems to perform secondary operations outside the mold. This reduces the cost of the mold, but increases the financial investment of performing the secondary operations, and increases the amount of processing and storage space requirements. Financial analysis will aid in making the final determination.

WHEN TO CONSIDER SECONDARY OPERATIONS

There are times when secondary operations are preferred to using mold design to eliminate the secondary operation requirements:

- *When volumes are small.* In the injection-molding industry, annual product volume of less than 25,000 pieces constitutes *low volume*. In some cases, this number may be as high as 50,000 pieces, but is still considered small. When a mold is built, it is understood that its cost must be absorbed in the selling price of the product. A common rule of thumb is that it is less expensive to perform secondary operations to a molded part than to increase the complexity, and cost, of the mold when the annual volumes are below 25,000 pieces. This assumes that the life of the product will not exceed 3 or 4 years.
- *When tooling costs are excessive.* Mold-building shops work under the same market demand concepts as any other manufacturing operation. This results in their prices fluctuating according to what the market will bear. There are times when the cost (and delivery time) of a mold may double simply because it is requested during one of the higher pricing phases of the economy. In such a case, the buyer may elect to reduce the amount of molded-in requirements and perform some secondary operations simply as a way of reducing the initial cost of the mold. Of course, the buyer could wait for the prices to come down, but this would undoubtedly result in a loss of market share or even a totally missed opportunity. In addition, most projects are funded in early budget-setting exercises which may dictate the total amount of money available for building a mold. The buyer will get everything possible built into that mold for the money available, but it may not be enough to cover all the requirements. In that case, some of the functions may have to be accounted for in secondary operations.
- *When time to build the mold jeopardizes marketing schedules.* Usually, the single most time-consuming part in the development of a molded product is building the injection molds. This can take from a few weeks for small, simple products to many months for complicated or large products. Normally, the amount of time allocated by early estimators for mold building is too short, and the result is a need to make the mold design-and-build phase much less complicated than may be desired, simply to meet marketing schedules. In that case, some of

the built-in features of the mold will have to be eliminated and the features accounted for through the implementation of secondary operations. This is a common problem, becoming more common as development cycles and expected product life cycles get shorter and shorter.

- *When a labor-heavy environment already exists.* There may be occasions when a company has too many workers on the payroll. It's possible that a large contract was unexpectedly canceled or marketing forecasts were negatively affected by economic conditions, or there may be a variety of other factors involved. The company may have spent a large amount of money and time training these employees and, believing that it will need the same employees in a short period of time, may wish to keep them rather than lay them off and go through the hiring and training process all over again. At times like these, it may be more economically advantageous to utilize the retained employees to perform secondary operations, thus reducing the initial mold investment and delivery time and maintaining a level payroll.

ASSEMBLY OPERATIONS

A variety of assembly operations exist, including snap fitting and use of screws and bolts. The following are some of the more common thermal and mechanical methods of assembling that go beyond the snap fits and screws. The section on ultrasonic welding details the variables that affect sonic welding, but these same variables generally apply to the other forms of welding as well.

Ultrasonic Welding

Ultrasonic welding is an assembly process that uses high-frequency mechanical vibrations (20,000 to 40,000 cycles [20 to 40 kHz] per second) transmitted through thermoplastic parts (thermoset materials cannot be ultrasonically welded in the traditional sense). These vibrations generate friction between the plastic parts being assembled, and this friction causes heat. The heat that's generated causes the mating plastic surfaces to melt slightly and fuse together, resulting in a welded product. Ultrasonic welding can be used for staking, surface (vibration) welding, spot welding, and inserting metal inserts. A typical ultrasonic welding machine is shown in Figure 9-1.

Both amorphous and crystalline materials can be ultrasonically welded, but crystalline materials require greater amounts of energy and are much

Figure 9-1. Ultrasonic welding machine. (Courtesy Branson Ultrasonics Corp.)

more sensitive to joint design, horn design, and fixturing. Basically, the higher the melt temperature, the more ultrasonic energy required for welding. The major factors affecting weldability, besides polymer structure (amorphous versus crystalline), are melt temperature, melt index (flow rate), material stiffness, and the chemical makeup of the plastic. Depending on compatibility regarding the factors just mentioned, certain dissimilar amorphous material combinations can successfully be welded together.

Energy Directors

Figure 9-2 shows a standard formula for determining the shape and size of a common *energy director* for sonic welding. The primary purpose of the energy director is to direct energy from the horn of the ultrasonic machine to the desired point of welding on the plastic part. The energy director focuses the ultrasonic energy to that point, thereby creating highly concentrated sonic energy, which causes the plastic to heat up quickly. Without the energy director, the sonic energy would be dissipated over the entire mating surface area and be very weak, causing longer cycle times due to the slower heating of the plastic.

Ease of Welding

Tables IX-1 and IX-2 indicate the relative ease of welding for some common thermoplastics. Ease of welding is a function of joint design, part geometry, energy requirements, amplitude, and fixturing. In addition, these ratings are based on *near field* welding (the welding joint is located within 1/4 in. [0.64 cm] of the horn contact surface).

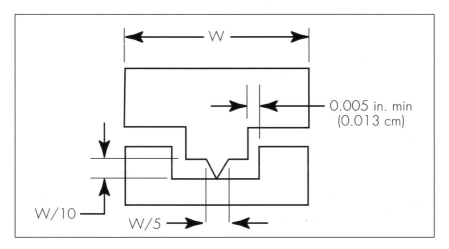

Figure 9-2. Energy director formula.

Table IX-1. Ease of Welding (amorphous materials)

Material	Standard Welding	Staking	Insertion	Spot Welding	Vibration Welding
ABS	E	E	E	E	E
ABS/polycarbonate blend	E-G	G	E-G	G	E
Acrylic (noncast)	G	F	G	G	E
Acrylic multipolymer	G	G	G	G	E
Butadiene-styrene	G	G	G	G	G
Polyphenylene oxide (PPO)	G	G-E	E	G	E-F
Polyamide-imide	G	N/R	N/R	N/R	G
Polyarylate	G	N/R	N/R	N/R	N/R
Polycarbonate	G	G-F	G	G	E
Polyetherimide	G	N/R	N/R	N/R	N/R
Polyethersulfone	G	N/R	N/R	N/R	N/R
Polystyrene (gen purpose)	E	F	G-E	F	E
Polystyrene (rubber mod)	G	E	E	E	E
Polysulfone	G	G-F	G	F	E
PVC (rigid)	F-P	G	E	G-F	G
SAN-NAS-ASA	E	F	G	G-F	E
PBT/polycarbonate blend	G	F	G	G	E
Code: E = Excellent, G = Good, F = Fair, P = Poor, N/R = Not recommended					

Table IX-2. Ease of Welding (crystalline materials)

Material	Standard Welding	Staking	Insertion	Spot Welding	Vibration Welding
Acetal	G	G-F	G	F	E
Cellulosics	F-P	G	E	F-P	E
Fluoropolymers	P	N/R	N/R	N/R	F
Ionomer	F	N/R	N/R	N/R	N/R
Liquid crystal polymers (LCP)	F	G-F	N/R	N/R	N/R
Nylon	G	G-F	G	F	E
Polyethylene terephthalate (PET)	G-F	N/R	N/R	N/R	N/R
Polybutylene terephthalate (PBT)	N/R	N/R	N/R	N/R	N/R
Polyetheretherketone (PEEK)	F	N/R	N/R	N/R	G
Polyethylene	F-P	G-F	G	G	G-F
Polymethylpentene	F	G-F	E	G	E
Polyphenylene sulfide (PPS)	G	P	G	F	G
Polypropylene	F	E	G	E	E

Code: E = Excellent, G = Good, F = Fair, P = Poor, N/R = Not recommended

The frequency for ultrasonic welding is usually 20 kHz (although 20 percent of welding jobs may require as high as 40 kHz). Vibration welding, on the other hand, is normally performed at low frequencies, on the order of 250 to 300 Hz. Vibration welding is normally required for large components, such as automotive bumpers, or intricate units that may be damaged by higher frequencies. Hermetic sealing usually requires vibration welding, also.

*Variables that Influence Ultrasonic Weldability**

Polymer structure. Amorphous resins are characterized by a random molecular arrangement and a broad softening temperature range (glass transition temperature T_g) that allows the material to soften gradually and flow without premature solidification. These resins generally are very efficient with regard to their ability to transmit ultrasonic vibrations and can be welded under a wide range of force-amplitude combinations.

---

*Information provided by Branson Ultrasonics Corp.

Crystalline resins are characterized by regions of orderly molecular arrangement and a sharp melting temperature T_m and resolidification point. The molecules of the resin in the solid state are springlike and internally absorb a percentage of the high-frequency mechanical vibrations of the ultrasonic generator. This makes it more difficult to transmit the ultrasonic energy to the joint interface, thus higher amplitude is usually required.

Melt temperature. Generally speaking, the higher the melt temperature of a resin, the more ultrasonic energy required for welding.

Stiffness (modulus of elasticity). The stiffness of the resin to be welded can influence its ability to transmit ultrasonic energy to the joint interface. Generally the stiffer the material, the better its transmission capability.

Moisture content. Some materials, such as nylon, ABS, polycarbonate, and polysulfone, are hygroscopic; that is, they absorb moisture from the atmosphere, which can seriously affect weld quality.

If hygroscopic parts are not sufficiently dried, when they are welded the moisture will become steam; this trapped gas will create porosity (foamy condition) and often degrade the resin at the joint interface. This results in difficulty in obtaining a hermetic seal, poor appearance (frostiness), degradation, and reduced weld strength. For these reasons, it is suggested that, if possible, hygroscopic parts be welded directly from the molding machine to ensure repeatable results. If welding can't be done immediately, parts should be kept dry by sealing them in polyethylene bags directly after molding. Drying the parts prior to welding can then be done in special ovens; however, special care must be taken to avoid material degradation.

Flow rates. Flow rate is the rate at which a material flows when it becomes molten, as indicated by the value obtained by the melt index test (ASTM D-1238). Different grades of the same material may have different flow rates. Such differences may result in the melting of one component of an assembly and not the other. Thus a melt or flow is created, but not a solid bond. Consult the resin specifications to ensure compatibility of flow rates.

Mold release agents. Often called *parting agents*, these modifiers are applied to the surface of the mold cavity to provide a release coating that facilitates removal of the parts. Agents such as carnauba wax, zinc stearate, aluminum stearate, fluorocarbons, and silicones can be transferred to the joint interface and interfere with surface heat generation and fusion, inhibiting welding. If it is absolutely necessary to use a release agent, the *paintable* or *printable* grades should be used. These will cause the least amount of interference with ultrasonic assembly.

Plasticizers. Plasticizers are high-temperature boiling organic liquids or low-temperature melting solids added to resins to impart flexibility. They do this by reducing the intermolecular attractive forces of the polymer matrix. They can also interfere with a resin's ability to transmit vibratory energy. Attempting to transmit ultrasonic vibrations through a highly plasticized material (such as vinyl) is like transmitting energy through a sponge. The energy is absorbed rather than directed to a focal point.

Flame retardants. Flame retardants are added to a resin to inhibit ignition or modify the burning characteristics. These chemicals are generally inorganic oxides or halogenated organic elements, and for the most part are not weldable. Typical retardants are aluminum, antimony, boron, chlorine, bromine, sulfur, nitrogen, and phosphorus. The amount of flame-retardant material required to meet certain test requirements may vary from 1 or 2 percent to 50 percent or more by weight of the total matrix, and the amount of available weldable material is reduced accordingly. This reduction must be compensated by modifying the joint configuration to increase the amount of weldable material at the joint interface and by increasing ultrasonic energy levels.

Regrind. Control over the volume and quality of regrind is necessary, as it may adversely affect the welding characteristics of the molded part with its potentially lower melting temperature. In some cases, 100 percent virgin material may have to be used to obtain the desired results.

Colorants. Although most colorants, either pigments or dyestuffs, do not usually interfere with ultrasonic assembly, a pigment loading of more than 5 percent can inhibit weldability. An application evaluation should be performed to determine feasibility and welding parameters for various pigments and loadings.

Resin grade. Welding different grades of the same resin can be difficult because of differences in melt temperature and molecular weight between the grades. Generally, however, both materials to be welded should have similar molecular weight and melt temperatures should be within 40° F (22° C) of each other.

Fillers. Fillers actually enhance the ability of some resins to transmit ultrasonic energy by imparting higher rigidity (stiffness). Common fillers such as calcium carbonate, kaolin, talc, alumina trihydrate, organic fillers, silica, calcium metasilicate, and micas can increase the weldability of the resin considerably with loadings of up to 35 percent. Above that level, there may not be enough resin at the joint surface to obtain reliable hermetic seals.

Even with lower percentage loadings, abrasive fillers can cause excessive wear on the surface of the ultrasonic horn. In this situation, the use of hardened steel or carbide-faced (coated) titanium horns is recommended.

Reinforcements. The addition of continuous or chopped fibers of glass, aramid, carbon, etc., can improve the weldability of a resin; however, rules governing the use of fillers should be observed. Long fibers can collect and cluster at the gate area during molding, being forced through in lumps rather than uniformly dispersed. This agglomeration can lead to a localized energy director containing a much higher percentage of reinforcement material in the resin. If this occurs, no appreciable weld strength can be achieved since the energy director would simply imbed itself into the adjoining surface, not providing the required amount of molten resin to cover the joint area. This problem can be minimized or eliminated by using short fiber reinforcements.

Hot-gas Welding

Rigid polyvinyl chloride (PVC), or simply vinyl, is the most common material assembled by the hot-gas welding method shown in Figure 9-3. Other materials that can be hot-gas welded are acrylonitrile-butadiene-styrene (ABS), ABS blends, acrylics, polyethylene, polypropylene, polystyrene, and polycarbonate. Normally, filled materials are not acceptable for hot-gas welding, but reinforced versions enjoy some success.

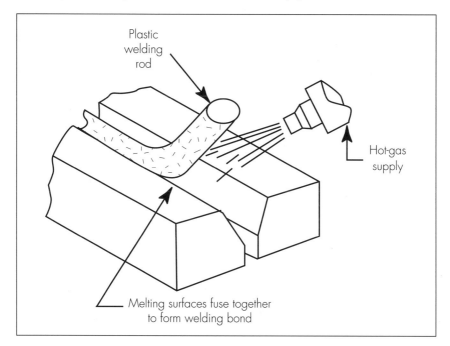

Figure 9-3. Hot-gas welding process.

The process is similar to metal welding in that a welding rod composed of the same material as that being welded is placed along a beveled joint area. Heat is then applied to that area by a hot gas, usually air, but nitrogen is recommended. The hot gas (40 to 1000° F [4 to 538° C]) melts the plastic to be joined as well as the welding rod. The gas tool continues on for further welding, and the plastic material in the heated area cools to resolidify and form a strong welded bond.

Induction (Electromagnetic) Bonding

The induction-welding process consists of activating an electrodynamic field to excite a conductive bonding agent (such as metal screening or wire strands), thereby creating heat in the agent. This heat is absorbed by the plastic components that surround the bonding agent, causing the plastics to melt. The melting plastics fuse together and to the bonding agent, and solidify once the electrodynamic field is deactivated. Figure 9-4 depicts this process. A slight pressure is usually applied to the components being welded. The total process takes between 1 and 10 seconds depending on the size of the area being welded.

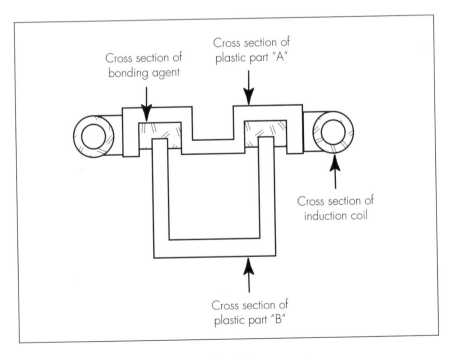

Figure 9-4. Induction (electromagnetic) welding concept.

Dissimilar materials can be welded together if they are thermally compatible—that is, their respective melting points are within 40° F (22° C).

A disadvantage of this process is that the conductive bonding agent remains sealed within the final welded part and is integral to that part. Therefore, a fresh bonding agent must be used for each welded assembly.

Spin (Friction) Welding

In the spin-welding process, two parts are brought together with one of the parts spinning at a speed between 100 and 1000 revolutions per minute (rpm) (Figure 9-5). Slight pressure is applied as the two parts are brought together. Special equipment can be built to perform this operation to allow complete control of parameters such as rpm, pressure, contact speed, and dwell. However, it may be possible to produce adequate welds with common shop equipment such as a drill press or lathe.

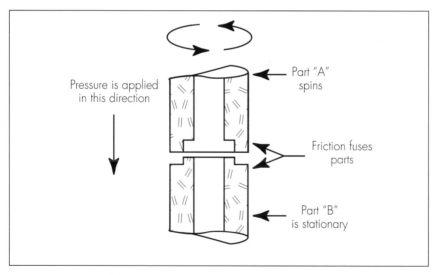

Figure 9-5. Spin welding.

Typical cycle times are in the range of 1 to 2 seconds. Although most rigid thermoplastics can be welded by this process, the softer materials such as low-density polyethylene create control problems and the welded area may be spongy and weak. A disadvantage of the spin-welding process is that normally the parts must be cylindrical. However, it is possible to design a circular welding ring on noncircular parts and align the parts properly to create friction on the circular ring.

Adhesive Bonding

Thermoplastic materials may be bonded with monomers and solvents, while thermoset materials require elastomerics and epoxies for bonding. Both require careful attention to joint design (Figure 9-6). Table IX-3 shows the adhesives best suited for some of the more popular plastic materials.

Monomer cements contain a specific plastic material that must be catalyzed to produce a bond. This can be done through heat, liquid catalyst, or ultraviolet (UV) light.

Solvent cements actually attack the surface of the plastic material in a controlled fashion and dissolve it, causing a molecular interlocking, after which the solvent evaporates.

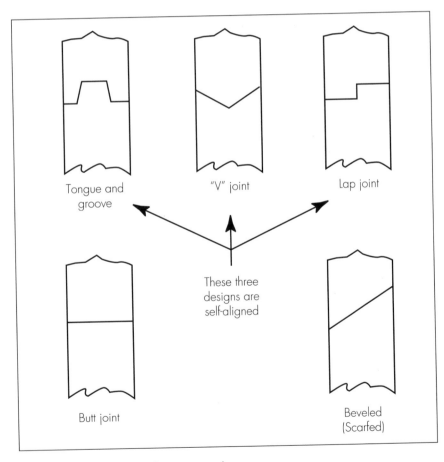

Figure 9-6. Adhesive bond joint examples.

Table IX-3. Adhesive Bonding Materials Compatibility

Plastic Resin	Adhesive Type
Acetals	1, 2, 10, 22, 27, 30
ABS	1, 2, 3, 8, 10, 12, 18, 21, 22, 24, 27, 30
Cellulose acetate	1, 2
Epoxy	2, 5, 8, 10, 20, 21, 22, 24, 26, 27, 30
Melamine	4, 6, 8, 10, 16, 27, 30, 32
Nylon	1, 2, 6, 7, 10, 12, 17, 22, 24, 30
Phenolic	2, 5, 6, 8, 9, 10, 17, 20, 21, 22, 23, 24, 26, 27, 28, 29, 31, 32
Polycarbonate	1, 2, 3, 8, 10, 12, 24
Polyester (TS)	2, 3, 8, 10, 18, 20, 21, 22, 24, 27, 30
Polyethylene	8, 10, 17, 22, 24, 27, 29
Polyethylene terephthalate (PET)	2, 5, 9, 16, 17, 21, 22, 26
Polyimide	8, 9, 10, 11, 22, 25, 26, 27, 28, 29
Polyphylene oxide (PPO)	2, 3, 8, 10, 21, 22, 24, 27, 30
Polypropylene	10, 22
Polystyrene	1, 3, 12, 17, 18
Polysulfone	22
Polyurethane	10, 22, 24, 27
Polyvinyl chloride (PVC)	1, 2, 8, 10, 12, 16, 18, 21, 22, 24, 27, 30
Silicone	2, 25, 26

Adhesive Code

Thermosets
1. Cyanoacrylate
2. Polyester + isocyanate
3. Polyester + monomer
4. Urea formaldehyde
5. Melamine formaldehyde
6. Resorcinol formaldehyde
7. Phenol formaldehyde
8. Epoxy + polyamine
9. Epoxy + polyanhydride
10. Epoxy + polyamide
11. Polyimide
12. Acrylate acid ester

Thermoplastics
13. Cellulose acetate
14. Cellulose acetate butyrate
15. Cellulose nitrate
16. Polyvinyl acetate
17. Polyamide
18. Acrylic

Elastomers
19. Natural rubber
20. Butyl
21. Polyisobutylene
22. Nitrile
23. Styrene butadiene
24. Polyurethane
25. RTV silicone
26. Silicone resin
27. Neoprene

Alloys
28. Epoxy phenolic
29. Epoxy nylon
30. Phenolic neoprene
31. Phenolic polyvinyl butyral
32. Phenolic polyvinyl formaldehyde

Elastomeric adhesives contain natural synthetic rubber in a water- or solvent-based solution. They cure at room temperature (faster at higher temperatures) and attain full bonding when the solvent or water is evaporated.

Epoxy adhesives (and similar-acting polyester and phenolic adhesives) act by forming a thermosetting layer of material between the two plastic surfaces to be bonded. This layer is catalyzed so that it cures and bonds with the base layers.

Note: Polypropylene, polyethylene, and the fluorocarbons are extremely difficult to bond adhesively because of their superior chemical resistance. It is considered better to bond these materials mechanically.

Joint design is instrumental in effecting a proper adhesive bond. Joints that combine both shear and tensile strengths are preferred. Although butt joints are sometimes successful, other designs such as lap joints, V joints, and tongue-and-groove joints should be used to ensure proper bonding of all plastics. These are depicted in Figure 9-6.

MACHINING OPERATIONS

Machining operations include drilling, tapping, lathe turning, milling, sanding, and to some extent, grinding. Any secondary machining process removes some of the surface material of the plastic product being machined, and therefore reduces the hardness and luster of that surface. This must be taken into consideration, especially for products that will be exposed to severe environments or that require a high luster for aesthetic purposes. In thermosets, and some thermoplastics, the luster may be restored by polishing the surface with mineral oil.

In most cases, plastics can be machined accurately if the proper tools, speeds, feeds, and coolants are employed. Thermoplastic materials are especially sensitive to cutting speeds and cutter shape because of the heat generated by the machining process. If excessive, this heat will cause deformation, or even melting, of the product being machined.

Drilling and Tapping Thermoplastics

Carbide drills are most suitable for drilling thermoplastics, but if they are not readily available, carbide-tipped or diamond-tipped drills can be used. If surface finish is critical and a mirror finish is required, use of a diamond-tipped drill is mandatory.

Flutes should be highly polished and the drill cutting surfaces should be chrome-plated or nitrided to reduce wear and increase cutting efficiency. Details of drill dimensions are shown in Figure 9-7. The drill land (L) should be 1/16 in. (0.16 cm) or less. The helix angle (HA) should be 30 to 40°. The point angle (PA) should be 60 to 90° for small drills (up to 1/8-in. [0.32-cm] diameter) and 90 to 115° for large drills (over 1/8-in. diameter). And the lip clearance angle (LCA) should be 12 to 18°.

Drill feed should be approximately 0.005 in. (0.013 cm) per revolution of drill bit. Drill speeds should range from 5000 rpm for drill diameters up to 1/8 in. to 1000 rpm for drill diameters of 1/2 in. (1.3 cm) or greater, with

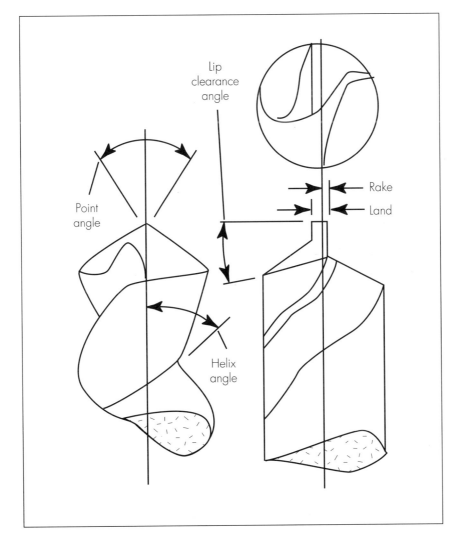

Figure 9-7. Drilling point details.

proportional speeds for drills between these diameters. Slower speeds may be required on certain materials such as PVC to prevent overheating.

Taps for thermoplastics should have two or three flutes and be made of solid carbide steel or chrome-plated (or nitrided) high-speed steel. Because of the resilience of some of the less-rigid plastics, it may be necessary to use a slightly oversized tap (0.001 to 0.005 in. [0.003 to 0.013 cm] over nominal) to ensure proper sizing and thread depth after machining.

Drilling and Tapping Thermosets

Because of the inherent abrasive nature of thermoset materials, it is recommended that carbide drills be utilized for drilling operations. Carbide-tipped or diamond-tipped drills are also acceptable. Flutes should be highly polished and the drill cutting surfaces should be chrome-plated or nitrided to reduce wear, reduce friction, and increase cutting efficiency.

The drill land (L) should be 1/16 in. (0.16 cm) or less (refer to Figure 9-7 for visual definitions of details). The helix angle (HA) should be between 15 and 30°. The rake (R) should be 0 to 3°. The point angle (PA) should be 90 to 115°. And the lip clearance angle (LCA) should be 12 to 18°. Because of the abrasive nature of the plastic materials, it is advisable to use drills that are 0.001 to 0.002 in. (0.003 to 0.005 cm) over size.

Drill feed should be approximately 0.005 in. (0.013 cm) per revolution of the drill bit. Drill speeds should range from 5000 rpm for drill diameters up to 3/32 in. (0.24 cm) to 750 rpm for drill diameters of 1/2 in. (1.27 cm) or greater, with proportional speeds for drills between those diameters. Higher speeds will result in improved finishes but will reduce drill life because of wear.

Taps for thermosets should have two or three flutes and be made from solid carbide steel or from chrome-plated (or nitrided) high-speed steel. Again, because of the abrasiveness of the plastic, it is recommended that 0.001- to 0.003-in. (0.003- to 0.008-cm) oversize taps be used.

Reaming Thermoplastics and Thermosets

Both thermoplastic and thermoset products can be reamed for accurate sizing of hole diameters. Reamers should be fluted for best surface finish and should be of carbide steel to minimize wear. For thermoplastic materials, it is best to use reamers that are 0.001- to 0.002-in. (0.003 to 0.005 cm) over size because of the resilience of the material.

Reamer feeds and speeds should approximate those of the drilling operations mentioned earlier, and water-soluble coolants should be used to achieve the best surface finish and minimize the effects of heat generated by friction.

Turning and Milling (Thermoplastics and Thermosets)

Lathe and mill cutters should be of tungsten carbide or diamond-tipped with negative back rake and front clearance (see Figure 9-8). Optimum tool bit designs should incorporate an X angle of 8°, a Y angle of 10°, and a Z angle of 20°.

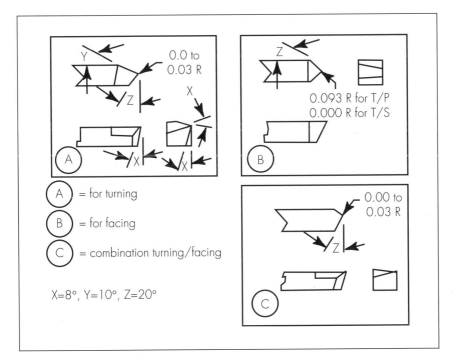

Figure 9-8. Design of turning points.

Feeds should be between 0.010 and 0.020 in. (0.03 and 0.05 cm) per revolution. Speeds should be between 200 and 700 ft/min (61 and 213 m/min) for thermoplastics and 1200 and 1800 ft/min (365 and 549 m/min) for thermosets. Use of water-soluble coolants will improve surface finish and reduce point wear.

AUTOMATED SHAPE CUTTING

Water Jet

Perhaps the most popular automated cutting process being used throughout all industries, water-jet cutting, employs the force of a thin stream of water under pressures in the range of 20,000 to 50,000 psi (137,890 to 344,720 kPa) to create a powerful cutting "point" that pierces plastic material cleanly and effortlessly. Dust and chips are nonexistent with water-jet cutting, and the addition of abrasive material to the stream allows cutting of the most difficult materials available today.

Although most water-jet cutting is performed on flat sheet stock, computer-controlled three- and five-axis machines are capable of cutting shapes on very complex surfaces. Some units combine water-jet cutting with more conventional mechanical cutting processes to provide a variety of machining operations at a single station.

Laser Cutting

Laser cutters are used when a fine polished finish on the plastic is required, such as on the edges of an acrylic sign. The laser unit cuts by focusing its concentrated beam at the exact point of the cut, which causes the plastic to melt, vaporize, and solidify, thus producing an ultrasmooth finish. Laser cutting can be compared to laser printing on paper. Instead of thermally leaving an inked imprint, the laser cutter leaves a freshly melted and solidified plastic section.

Proper adjustments of laser energy and cutting speeds are critical to a successful cutting operation; improper adjustments can cause the plastic to char, burn, or disintegrate, resulting in the release of toxic fumes. Adequate exhaust venting is imperative to preclude this potential danger. Although most lasers operate effectively in the power range of 200 to 500 W, some plastics require higher-power cutters. Equipment is now available with lasers operating at 1000 W and higher.

SURFACE FINISHES AND DECORATING PROCEDURES

Preparation of Surface

Products that require postmold decorating require surface preparation to ensure adequate bonding of the decorative material to the plastic material. This surface preparation can range from a simple detergent wash to a complex acid bath, depending on the type of plastic being decorated and the decoration process. Some of the typical surface treatments beyond simple detergent washes follow.

Flame Treatment

This is the most common method of preparing polyolefin and acetal plastics for decorating. These plastics, being crystalline materials, are highly resistant to chemicals, paints, and inks because of the slippery nature of their molded surfaces. Flame treatment, which consists of passing the molded product through a flame, causes the surface of the plastic to oxidize, making it receptive to adhesion of paints, inks, and other decorating media. The procedure requires a fair degree of control to ensure that the surface is oxidized properly without degrading or charring the plastic, but with practice, the process can be done well by hand, or automated for closer control of parameters.

Corona Discharge

Surface oxidation of plastic material can also be achieved through the use of a corona discharge process in which the plastic product is passed over an insulated metal cylinder beneath a high-voltage conductor. An electron discharge then takes place between the conductor and the cylinder and strikes the surface of the plastic product passing through. A corona is formed that causes the plastic to oxidize on the surface, thus preparing it for decorative coatings.

Plasma Process

In this process, low-pressure air is directed through an electrical discharge and expanded into a vacuum chamber containing the plastic product to be treated. While it passes through this chamber, nitrogen and oxygen gases are partially disassociated from the air and, in their atomic state, react with the surface of the plastic to alter the physical and chemical characteristics. This altered surface then readily accepts decorative coatings.

Acid Etch

Some plastics, such as polycarbonate and some grades of ABS, do not accept decorative finishes even after exposure to treatments such as those mentioned above. In these cases, it may be necessary (or preferable) to obtain a mechanical bond between the plastic and the decorative coatings. This can be done through use of an acid wash process that attacks the surface of the plastic and creates microscopic craters of exposed resin. These craters will then physically capture the decorative coating and lock it to the plastic surface.

Applied Finishes

Painting

Paint can be applied to a plastic product by way of brushing, spraying, rolling, or dipping, either manually or mechanically. Most plastics are painted using the standard spray process (Figure 9-9). Spray painting is referred to as a *line-of-sight* process, meaning that what is painted is only what can be seen by the spray gun unit.

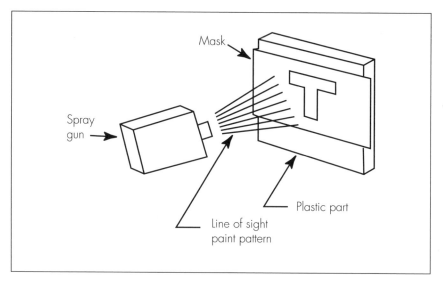

Figure 9-9. Spray painting.

Although most plastics can be painted, the success of the process depends on proper surface preparation and the use of specific paint formulations for specific plastics. These formulas must be adjusted to address (1) the plasticizer (if any) in the plastic to make sure there is no migration into the paint; (2) the heat distortion temperature of the plastic, which determines whether the paint should be air dried or oven baked; and (3) the chemical resistance properties of the plastic to determine which solvent system should be used to ensure proper adhesion without crazing. In some cases, water-soluble paints have found successful applications.

Plating (Electroplating)

Electroplating plastic material requires that the normally nonconductive plastic be made conductive. That usually necessitates first applying a conductive base metal to an etched and sensitized plastic surface. Then a plating material can be applied to the conductive metal. Procedures vary, depending on the specific plastic being plated, and detailed plating information is available from suppliers of the specific plastic.

Metallic plating of plastics may be required for either decorative or functional purposes. Examples of decorative purposes are escutcheons, plumbing fixtures, and jewelry finishes. Examples of functional purposes are circuit board traces, electromagnetic interference (EMI) shields, and corrosion-resistant surfaces.

Not all plastics can be plated, and those that can usually require that a specific plating grade be used. Some of the more popular plastics that accept plating include ABS, acetal, acrylic, alkyd, cellulosics, epoxy, phenolic, polycarbonate, polyurethane, tetrafluoroethylene (TFE) fluorocarbon, urea, and rigid PVC.

Vacuum Metallizing (Deposition)

When a plastic product needs a bright metallic finish but does not need the thick, rigid, protective plate created by electroplating, the less expensive method of vacuum metallizing should be considered (Figure 9-10).

This deposition process requires that the plastic product be coated with a lacquer base coat to minimize surface defects and increase adhesion properties of the plastic substrate. The product is then placed on a rack inside a vacuum chamber, along with small clips of the metal to be deposited. During operation, the metal clips are electrically heated to the point of vaporizing, at which point, because of the high vacuum in the chamber,

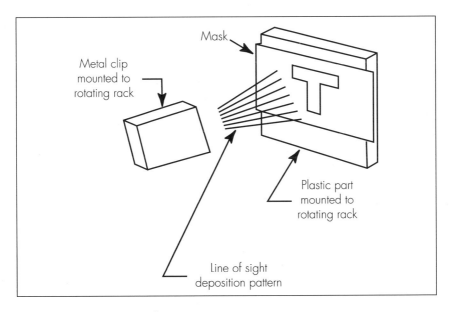

Figure 9-10. Vacuum metallizing process.

the metal vapor deposits on all line-of-sight surfaces. The product is then removed from the chamber and coated with another layer of lacquer to protect the metal finish, which is only about 5 millionths of an inch (5 μ in. [130 nm]) thick. If a color tone is required, it is applied at the same time as the final lacquer.

Hot Stamping*

There are three main methods of hot stamping, all done on a hot stamp machine (Figure 9-11): roll-on decorating, peripheral marking, and vertical stamping.

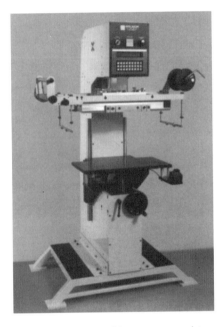

Figure 9-11. Typical hot stamp machine. (Courtesy United Silicone)

Roll-on decorating (Figure 9-12) is ideal for applying foils or preprinted heat transfers to part surfaces with large areas. With this method, a silicone rubber roller applies heat and pressure to release the print medium onto the plastic substrate. The advantage of this process is that the rubber roller material maintains line contact and pushes out trapped air between the printed medium and the decorating surface so that air bubbles are eliminated.

Peripheral marking (Figure 9-13) is mainly used for applying foils or preprinted heat transfers to the periphery of cylindrical or slightly conical parts. With this method, the plastic product is rolled under a flat stamping die, or roller, that applies heat and pressure to release the print medium onto the plastic substrate. The advantage of this process is that up to 360° of the part circumference can be decorated in one machine cycle.

Vertical stamping (Figure 9-14), the most common hot stamping method, is ideal for applying foils or preprinted heat transfers to smaller areas of flat or slightly crowned plastic products. It can also be used for up to 90°

---

*This information supplied by United Silicone, Inc., 4471 Walden Ave., Lancaster, New York, 14086.

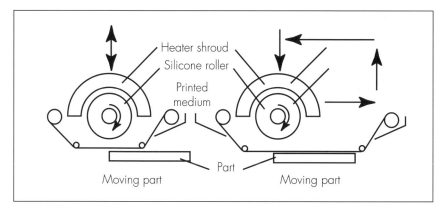

Figure 9-12. Common roll-on decorating method. (Courtesy United Silicone)

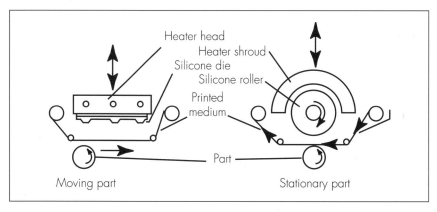

Figure 9-13. Common peripheral marking method. (Courtesy United Silicone)

on the circumference of cylindrical or spherical products. Typically, a silicone rubber die is mounted to the heated head of a vertical machine and positioned directly over the part to be decorated. The rubber die contains raised graphics (approximately 1/32 in. [0.08 cm]) to be stamped and is heated to the approximate melting point of the plastic to be stamped. Placed between the rubber die and the plastic product is the hot stamp foil, which consists of various thin coatings deposited on a film carrier. The rubber die is lowered and pushes the foil against the waiting plastic product. Heat from the die causes the release agents in the foil to activate and also softens the surface of the plastic product. The decorative resins in the foil are transferred and are thermally bonded to the plastic part.

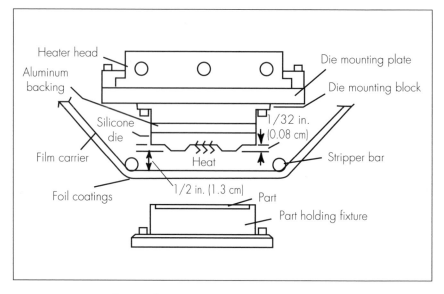

Figure 9-14. Typical vertical stamping method. (Courtesy United Silicone)

Pad Printing (Heat Transfer)*

Pad printing is usually done like printing paper on a press. A pad (usually made of silicone rubber) is inked with the image to be placed on the plastic product (Figure 9-15). This inking is performed by pressing the rubber pad onto a steel or nylon plate on which the image is etched or machined with ink screened into that image. The inked pad is then brought to the surface of the plastic product and pressed against it, transferring the inked image.

Screen Printing

In this process, ink or paint is forced through the mesh of a plastic or metal screen by pulling a squeegee across a screen that is placed against the surface of the plastic product being decorated (Figure 9-16). An artwork mask secured to the screen allows the ink to flow through the mesh only in specific areas, thus forming the required design on the product. The artwork mask is in the form of a stencil made by placing the artwork positive on a photosensitive film and exposing the film to a light source that etches the image onto the film.

---

*This information supplied by United Silicone, Inc., 4471 Walden Ave., Lancaster, New York 14086.

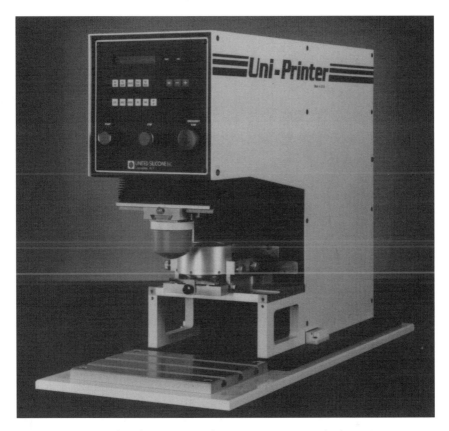

Figure 9-15. Typical pad printing machine. (Courtesy United Silicone)

In-process Finishes

Molded-in Color

An alternative to painting a color on plastic products is to mold the color into the plastic. The advantages are (1) the color does not have to be applied as a secondary operation and (2) the color does not wear off the surface during the product's life cycle.

Molded-in color is produced by blending a coloring agent with the plastic pellets used during the injection-molding process. Usually, this agent is a powdered dye or concentrate. Sometimes liquids are used.

Dry color is mixed with the raw plastic pellets prior to injection molding, usually by an automatic blender mounted on the molding machine hopper. Sometimes, however, the dry colorant can be tumbled in with the raw pellets by hand-mixing in a large container such as a clean metal barrel or cement mixer. Dry colorants are usually shipped in small bags that

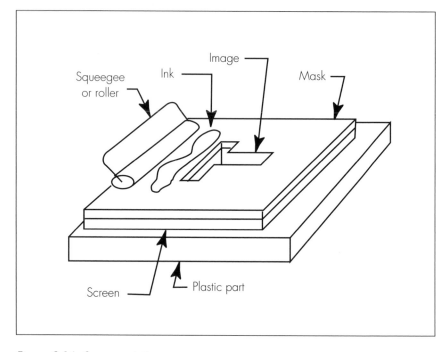

Figure 9-16. Screen printing process.

are premeasured to be mixed with 50-, 100-, or 200-lb (23-, 45-, or 91-kg) batches of raw pellets.

Color concentrates are produced by extruding heavy concentrations of colorant agent into a basic batch of plastic material compatible with the pellets to be colored. These heavy colored batches are then mixed with the basic pellets at approximately 5 percent ratio (5 lb concentrate to 100 lb pellets [2 kg concentrate to 45 kg pellets]), and the mixture is injection-molded as a total blend.

Molded-in Symbols

For ease of decorating, and to cut secondary costs, molded-in letters, symbols, and designs can be coated by using a rubber roller with ink or paint applied to its surface. The molded-in image is created by machining the image directly into the mold steel. This produces a raised image on the molded part, and this raised image receives a coating only on the top of the raised surfaces much as a rubber stamp against a stamp ink pad.

In some cases, at higher cost, the molded-in image can be embossed into the surface of the plastic product. For decorating these symbols, the ink or paint is wiped into the recessed image and then the surface is wiped clean, leaving only the coating that is in the recesses.

Two-color (Two-shot) Molding

Products such as typewriter keys and computer keyboard keys are susceptible to surface wear by constant use. For this reason, a process known as *two-color molding* is used for creating the decorative finish for keyboards and similar products. This process is actually an injection-molding process performed twice (Figure 9-17).

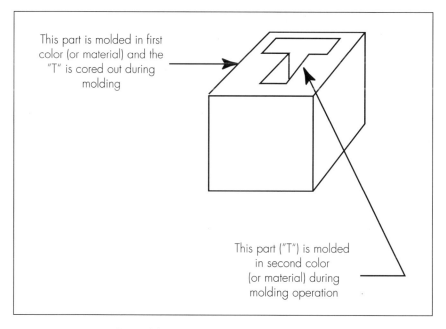

This part is molded in first color (or material) and the "T" is cored out during molding

This part ("T") is molded in second color (or material) during molding operation

Figure 9-17. Two-color molding concept.

In the first step, the base color material is molded into a basic shape. Then the second material is injection-molded into the remaining open spaces. In the case of a keyboard key, the key itself is molded in an off-white material, but the area where a symbol is intended is molded as an open space. That space is then filled during the second injection step with a material of a different color such as black. As the surfaces wear in use, they wear consistently so that the two colors are always apparent.

While two-color molding uses more expensive equipment than ink-stamping or painting, it is recommended when wear is a factor to be considered as a function of final product design.

Textured Surfaces

Textured surfaces can be painted onto a plastic surface, but in most cases they are molded directly into the product. This is accomplished by etching or machining the surface of the cavity image in the mold. The amount of texture depth and size is determined by preselecting a pattern from samples available from companies specializing in this type of work. One word of caution: when comparing samples, try to have the samples created in the same material and color that will be used in the molded product. A specific texture shown in a red ABS will take on a totally different appearance in a black nylon.

Textures are available in hundreds (if not thousands) of variations ranging from a fine satin finish to a heavy alligator hide. They are usually used for aesthetic purposes only, although some heavy textures may even add structural strength to the molded product. They are also used to hide molding imperfections such as splay, knit lines, and blush.

The biggest disadvantages of using textures are the high original cost to create the texture in the mold, the fact that the texture will eventually wear off the mold in high-friction areas and must be replaced, and the problem of selecting a texture based on previously molded samples that are not exactly like the final product to be produced.

In-mold Overlays

For thermoset products, in-mold overlay decorating is done by placing a foil (overlay) in the mold, consisting of a thermoplastic sheet that has the required image printed on it (Figure 9-18). This is then coated with a partially cured thermosetting resin. During the compression-molding process, the entire overlay is fused to the curing molding compound and becomes one with the finished product.

For thermoplastic products, a similar process is used. However, because the high pressures created during the injection process would simply tear apart the overlay, a slightly different approach is taken. This process, called *low-pressure molding*, incorporates injecting a charge of plastic material, under very low pressure, into an open mold. The mold is then closed and the plastic is compression-molded to create the final product. This keeps the fragile overlay from being destroyed.

A recent development in *controlled-pressure* molding* allows thermoplastic materials to be injected at high pressures, but slowly, so that overlays placed in the mold do not wrinkle or tear. This process reduces the injection pressure just before the mold cavities are filled, thus minimizing

---

*Patented by Hettinga Equipment, Inc., Des Moines, Iowa.

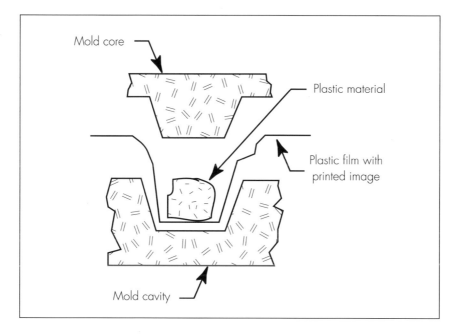

Figure 9-18. In-mold overlay process.

the amount of clamp force required. This means that molds can be run in presses with much smaller clamp units than those used for standard injection processes, and the molded products will exhibit much less molded-in stress because of the lower final injection pressures.

SUMMARY

A secondary operation can be defined as any operation performed on a product after it has been molded. Such operations normally include, but are not limited to, assembly, machining, and finishing (including decorative finishing).

Secondary operations should be considered when (1) the annual volume requirements are less than 25,000 pieces, (2) tool costs are inordinately high, (3) production schedules may be jeopardized by time to build sophisticated molds, and (4) a labor-heavy environment exists.

Ultrasonic bonding processes can be used to weld a wide variety of plastics. Crystalline materials require greater amounts of energy than amorphous and are much more sensitive to joint design, horn design, and fixturing. Basically, the higher the melt temperature, the more ultrasonic energy required for welding. The major factors affecting weldability include polymer structure (amorphous versus crystalline), melt temperature, melt index (flow rate), material stiffness, and the chemical makeup of the plastic being welded.

Surface and decorative finishes are dependent on proper surface preparation of the plastic product being finished. Usually, a specific grade of plastic must be used, and the surface must be prepared by using anything from a basic detergent wash to more sophisticated (and costly) etching processes.

There are two classifications of surface finishes: those applied during the molding processes are referred to as *in-process* finishes, and those applied after the molding process are referred to as *applied* finishes. In-process finishes include molded-in color, molded-in symbols, two-color molding, textured surfaces, and in-mold overlays. Applied finishes include painting, plating, vacuum metallizing, hot stamping, pad printing, and screen printing.

QUESTIONS

1. What is meant by a *secondary operation*?
2. List two circumstances in which secondary operations should be considered.
3. What is the normal frequency range of sound waves created by the sonic welding process?
4. What are two variables that influence ultrasonic weldability of plastic products?
5. What adhesive could be used for successfully bonding polysulfone to nylon?
6. What is the range of drill speeds recommended for drilling most thermoplastics?
7. Why is surface preparation usually required before a finish is applied to plastic parts?
8. List two common surface preparation treatments.
9. Define the difference between *applied* finishes and *in-process* finishes.
10. List three each of the more common applied finishes and in-process finishes and identify which is which.

Testing and Failure Analysis   10

OVERVIEW

The terms *testing* and *failure analysis*, as we use them here, should be defined separately. Testing is performed to analyze the basic material and product design concepts that are incorporated to manufacture a molded product. Failure analysis is activity performed on molded products that have failed to meet their intended design criteria, either shortly after being molded or in use by a consumer. We will look at both individually, although some of the same procedures and methods may be used in both situations. The procedures and methods discussed are not all-inclusive, only representative; they do not reflect exact methodology or principles. The source for most of the information that follows is the American Society for Testing and Materials (ASTM) book of standards.* In addition, we list the equivalent test standard number assigned by the International Organization for Standardization (ISO), which is becoming more accepted on a global basis. In a few cases, no ASTM or ISO tests are applicable.

TESTING

The type of testing we discuss requires calibrated equipment, well-documented procedures and results, proper training of test personnel, multiple test runs, and proper sample preparation. Proper preparation of the sample to be tested is paramount. Improper sample preparation can cause improper test results. The following represent some of the most common tests performed. There are others, of course, and specific products and applications will require performance of specific tests.

Electrical Testing

Plastics are good insulators. Because they are considered nonconductive, they are used for such products as screwdriver handles and connector

---

*Revised annually and available from American Society for Testing and Materials, 1916 Race St., Philadelphia, Pennsylvania 19103.

housings. These products depend on the ability of the plastic to withstand exposure to electrical current, and testing procedures have been developed to monitor the capabilities of the plastics to fulfill their mission. The common tests for electrical requirements are dielectric strength, dielectric constant, volume resistivity, surface resistivity, and arc resistance. These are explained in the following paragraphs.

Conditioning Samples

Electrical testing is dependent on the proper moisture level being present in the plastic sample to be tested. The sample must be conditioned, per ASTM test D-618, to establish this level. The basic conditioning method, known as Procedure A, consists of placing the sample in a standard laboratory environment of 73.4° F (23° C) and relative humidity of 50 percent for a minimum of 40 hours, for samples up to 0.250 in. (0.64 cm) thickness. For samples over this thickness, the time is increased to a minimum of 80 hours. The samples must be tested immediately if removed from this conditioning environment. Best results are attained when testing is performed while the samples are still in the environment.

Dielectric Strength—ASTM D-149 (ISO IEC 243-1)

This test is designed to measure the amount of voltage required to arc through a specimen of plastic (Figure 10-1). Voltage, starting at 0, is applied to one side of the specimen and increased until it arcs through the specimen. The specimen can be a sample cut directly from a molded product or a flat sheet representing the same material and thickness as a proposed product.

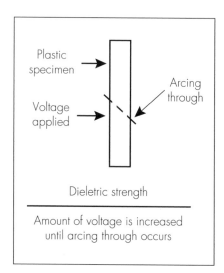

Figure 10-1. Dielectric strength.

Dielectric Constant—ASTM D-150 (ISO IEC 250)

The dielectric constant test is designed to measure the electrical capacitance of a specific plastic cross section as a ratio to that of a similar cross section of air (Figure 10-2). The frequency range that can be covered extends from less than 1 hz to several hundred megahertz.

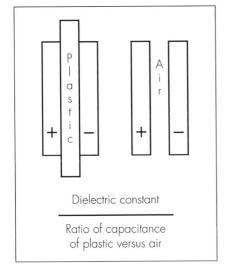

Figure 10-2. Dielectric constant.

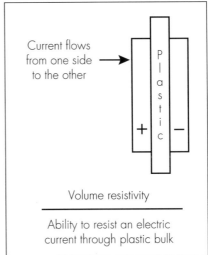

Figure 10-3. Volume resistivity.

Volume Resistivity—ASTM D-257 (ISO IEC 93)

This test is used to measure the ability of a plastic to resist an electric current through its bulk (Figure 10-3). The test is used as an aid in designing electrical insulators.

Surface Resistivity—ASTM D-257 (ISO IEC 93)

Similar to that used for determining volume resistivity, this test measures the ability of a plastic to resist current across its surface (Figure 10-4). This property is of importance in design of products such as circuit boards and connectors.

Arc Resistance—ASTM D-495

The purpose of this test is to measure the amount of time required for an electrical arc to carbonize the surface of a specific plastic specimen, thereby making it conductive (Figure 10-5). Although the results of this test are not generally used for specification purposes, they are valuable in the initial material selection process.

Physical Testing

Tests in this category determine the physical values of such properties as shrinkage rate, density (specific gravity), water absorption, moisture content, and melt flow index. These properties affect the ability of a specific

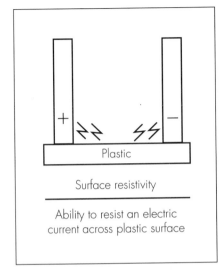

Figure 10-4. Surface resistivity.

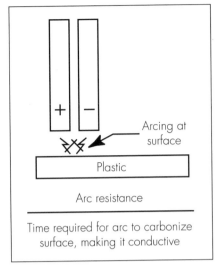

Figure 10-5. Arc resistance.

plastic to be processed during injection molding, and are instrumental in determining the final appearance of the molded product.

Shrinkage Rate—ASTM D-955 (ISO 294-4)

This test is used to measure the amount of shrinkage (in./in. [cm/cm]) that occurs in a specific plastic after it has been heated and injected into a mold, then allowed to cool (Figure 10-6). Initial measurements are taken between 2 and 4 hours after removal from the mold and additional measurements are taken at approximately 20 hours and 44 hours following removal from the mold.

The amount of shrinkage must be measured both parallel with and across the direction of flow. Two different instruments are used to make those measurements, as shown in Figure 10-6. In addition, there are several variables that affect shrinkage, such as the material temperature, flow rate, injection pressure, size of sprue and nozzle, percentage of reinforcement (if any), and other factors. For this reason, this test is to be used for reference purposes only.

Density—ASTM D-792 (ISO 1183:1987)

This test is used to determine the weight of a specific volume of a particular plastic. The measurement is stated as grams per cubic centimeter (g/cm^3) and will range from approximately 0.95 to 1.60. As a reference, water

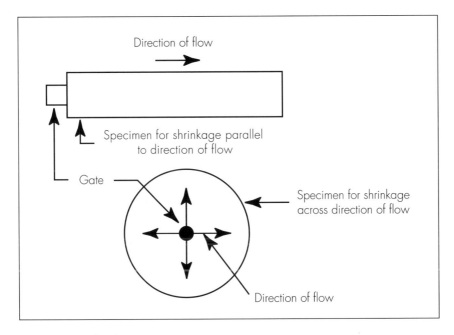

Figure 10-6. Shrinkage rate.

is considered to have a density of 1.0 g/cm³. A plastic with a value of less than 1.0 will float on water; one with a value of more than 1.0 will sink.

Density is actually a measurement of the weight of 1 cubic centimeter of a plastic part (or material) (Figure 10-7). Because of this, density is closely related to specific gravity, which compares the weight of a material in air to the weight of that same material in water. The common *water volume displacement test*, which is used for determining specific gravity, can also be used for determining density. Basically, this test measures the amount of water displaced by a given volume of plastic. First, the plastic piece is weighed in air by suspending it on a wire from a weighing device. Then the plastic piece, while still suspended on the wire, is weighed again while being placed in a container of water. A ratio is established that shows the weight of the object in air (A) divided by the weight of the object in air minus the weight of the object in water (B):

$$\text{Density} = \frac{A}{A\text{-}B}$$

Water Absorption—ASTM D-570 (ISO 62:1980)

This test determines the amount of moisture that is absorbed by a plastic material over a 24-hour period. Depending on the specific plastic being

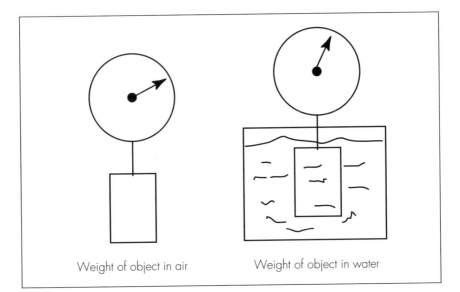

Weight of object in air   Weight of object in water

Figure 10-7. Density.

tested, however, the measurement period may be as much as a matter of weeks.

Properly sized and conditioned samples are weighed and placed in a container that allows them to be totally immersed in water (Figure 10-8). After 24 hours, the samples are removed, surface moisture is wiped off, and the samples are weighed again. The amount of increase is expressed as a percentage of weight increased by absorption of water. For materials that absorb high levels of moisture, the test can be repeated by immediately re-immersing the samples in water. Measurements may be taken every 24 hours until the amount of increase is less than 5 mg. At that point, the sample can be considered completely saturated.

Moisture Content—Tomasetti Volatile Indicator (TVI)

This simple test is named after the General Electric Plastics Section application engineer who developed the technique. It is an inexpensive, accurate method of determining the presence of moisture. Sometimes called the *resin dryness test*, the TVI test (Figure 10-9a) requires only a hot plate (capable of maintaining 525° F [274° C]) temperature, two glass laboratory slides, tweezers, and a wooden tongue depressor.

The test is performed by placing two glass slides flat on a hot plate. The hot plate is then turned on and set to maintain a temperature of 525° F. When the hot plate reaches and maintains that temperature, three or four

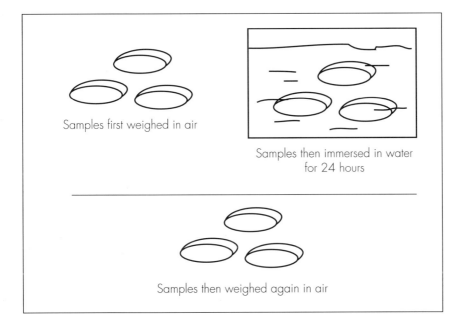

Samples first weighed in air

Samples then immersed in water for 24 hours

Samples then weighed again in air

Figure 10-8. Water absorption.

plastic pellets are deposited on one of the slides with the tweezers. The second slide is then immediately placed on top of the first, creating a sandwich of two slides with the pellets between them. The tongue depressor is then used to press the sandwich of slides and pellets until the pellets melt and flatten into 1/2- in. (1.3-cm) circular patterns.

Figure 10-9b shows typical results of the TVI test. The patterns on the left represent evidence of moisture. The moisture results in bubbles being formed in the melting resin. If only one or two bubbles is present, the indication is one of trapped air rather than moisture. The patterns on the right represent dry material with no moisture. This material would be acceptable to mold as is, if used within an hour or two.

Melt Flow Index—ASTM D-1238 (ISO 1133:1991)

This test is used to predetermine the processability of a specific plastic. It can also be conducted to establish the batch-to-batch consistency of material as it is provided by the material supplier.

The test is performed by placing material (regrind or virgin) in the preheated barrel of the machine (Figure 10-10). The barrel is heated to a specific temperature, depending on the plastic being tested. Then a weight is placed on the end of the plunger rod which causes the plunger to move

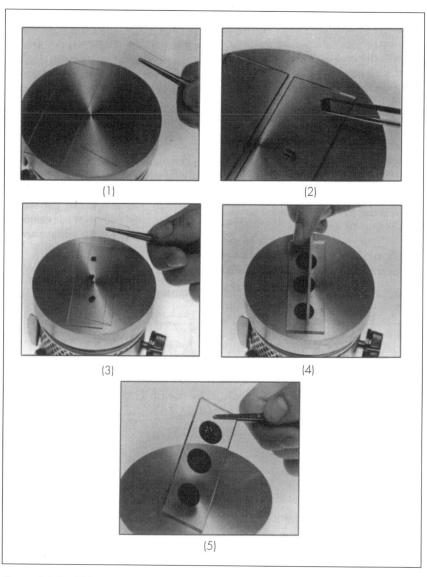

Figure 10-9a. TVI moisture content test. (Courtesy GE)

forward. The amount of plastic that exudes from the nozzle over a 10-minute period is measured. The test value is expressed as grams per 10 minutes, and will usually range between 4 and 20, depending on the flowability of the specific grade and family of plastic being tested.

The primary value of this test is as a quality control for incoming material. A value can be established as ideal for a specific product. That value can then be requested of the material supplier for all future material purchases, and a letter of certification can be requested of the supplier for each shipment. When the material arrives, it can be tested for melt index value. If the value is out of range, the material can be returned (at the supplier's expense), or used as long as it is understood that material properties may be affected. Table X-1 shows how properties may be altered by a value that is lower than requested. Note that permeability and gloss actually *decrease* as the melt index value drops.

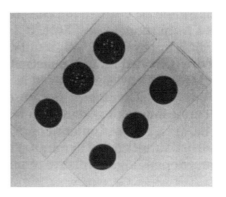

Figure 10-9b. TVI patterns on test slides. (Courtesy GE)

Mechanical Testing

Mechanical testing is performed to determine structural properties, such as tensile, compression, flexural, creep, and impact strengths, of a specific plastic material. These tests must be performed on specifically designed and processed samples and not on the actual product itself. As with most tests, the samples must be properly conditioned prior to being tested. Results

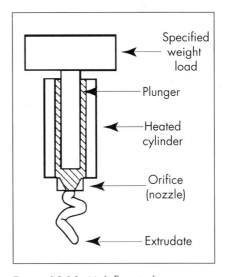

Figure 10-10. Melt flow index.

of the tests should be used for reference purposes only; they are also a valuable aid in selecting the proper plastic material for a specific design application.

Tensile Strength—ASTM D-638 (ISO 527-1 and 2:1993)

Tensile strength testing is performed to determine the point at which a plastic sample either breaks or yields. The measurement is useful in selecting materials that will be exposed to tensile-type (pulling) actions.

Table X-1. Melt Index Value's Impact on Plastics Properties

As melt index value *decreases*:

Stiffness	Increases
Tensile strength	Increases
Yield strength	Increases
Hardness	Increases
Creep resistance	Increases
Toughness	Increases
Softening temperature	Increases
Stress-crack resistance	Increases
Chemical resistance	Increases
Molecular weight	Increases
Permeability	Decreases
Gloss	Decreases

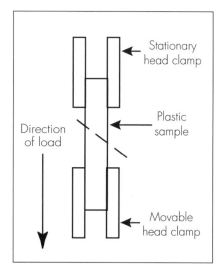

Figure 10-11. Tensile strength testing determines break and yield points in plastic materials.

The test is performed by gripping a sized and conditioned sample in a machine with a movable head and a stationary head (Figure 10-11). The moving head is activated and the sample is slowly pulled apart. Measurements are taken at the point of breakage or yielding. These are expressed as pounds per square inch (pascals).

Compressive Strength—ASTM D-695

Although compressive strength testing is seldom utilized, it does aid in determining the ability of a plastic to withstand the application of compression forces. This data may be useful in designing products such as those used for supporting shelves or overhead structures. Under normal conditions, flexural strength and creep resistance testing are preferable to compression testing.

The test is similar to tensile testing, but instead of pulling the sample apart, this test pushes it together until it breaks or yields (Figure 10-12).

Flexural Strength—ASTM D-790
(ISO 178:1993)

Flexural tests are related to the stiffness of the plastic and its resistance to bending. The test begins by placing a flat specimen across two beams. An opposing load is then applied to the specimen until the specimen either breaks or yields (Figure 10-13).

Creep—ASTM D-674

While creep (or deformation under load) information may be valuable for designing products that will support a load over along period of time, there is no established method of determining creep. The ASTM has provided a method that is frequently used, but because of the complications involved with measuring creep, even this test should be used only for research purposes and not definitive, routine testing.

The test consists of clamping a specimen at one end and hanging a load on the other end (Figure 10-14). The amount of stretching that takes place is measured first hourly, then daily, then weekly, and finally (if required) monthly, for a total time period of up to a year. Any measured stretching (creep) is plotted on a graph that shows the tendency and rate of creep for a specific plastic.

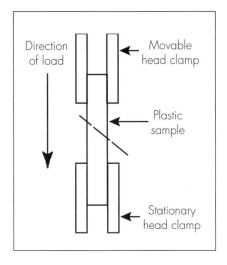

Figure 10-12. Compression testing.

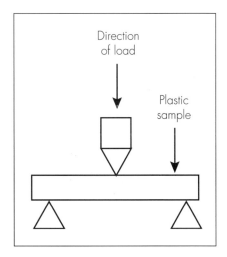

Figure 10-13. Flexural strength testing.

However, there are so many variables that come into play with this test that it cannot be used for anything but reference testing.

Impact Testing—ASTM D-256 (ISO 179 and 180:1993)

Two basic test methods are used for determining impact strength. These are known as *Izod* (vertical beam) and *Charpy* (horizontal beam) tests. These

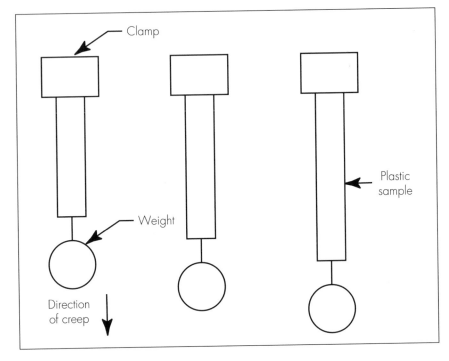

Figure 10-14. Creep testing.

tests are performed on notched or unnotched samples. Figure 10-15 illustrates the notched method. For both tests, a recording is made of the travel of the impact head (pendulum), both with and without a specimen mounted. The difference is calculated as an energy loss by the pendulum, and is referred to as the *impact resistance* of the plastic.

Thermal Testing

Five basic characteristics are determined in tests of thermal properties of plastic materials: melting point, heat deflection temperature, Vicat softening temperature, flammability, and oxygen index. These tests are performed to determine end-use properties as well as processing parameters.

Melting Point—ASTM D-3418 (ISO 3146:1985)

This temperature is referred to as the melting temperature (T_m) for crystalline materials and the glass transition (T_g) for amorphous materials. It can be thought of as the temperature at which the plastic material is readily flowing and able to be properly injected. It can be used to determine the

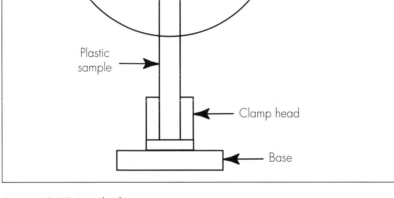

Figure 10-15. Notched impact testing.

starting temperatures at which the injection barrel of the machine can be set for initial processing.

The test is commonly performed using a Differential Scanning Calorimeter. This machine measures the temperature difference, and energy necessary to establish a "zero" temperature difference, between a specimen and a reference sample. It records this data as a curve. In Figure 10-16a, this curve is dramatic and shows a definite peak. The tip of this peak is the temperature at which the crystalline material reached the melting temperature. The curve in Figure 10-16b does not have a sharp peak because the material is not crystalline, but amorphous. However, by magnifying the area in which the glass transition point should be found, we see a pattern on the line that resembles an "S." By drawing a line through the curves of this "S," we can determine the exact temperature at which the amorphous material goes through the glass transition.

Heat Deflection Temperature (HDT)—ASTM D-648 (ISO 75-1 and -2:1993)

The heat deflection temperature is a good reference point for determining the temperature at which molecular action takes place and the plastic

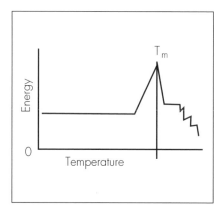
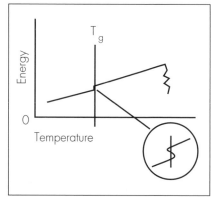

Figure 10-16a. Crystalline melting temperature graph, exhibiting its recognizable peak.

Figure 10-16b. Amorphous glass transition has no peak, but is recognizable by its s-shaped transition point.

material can flow. It should *not* be used to determine the end-use temperature limits of a specific product design. In fact, it is not a practical test for any reason because it is performed under a load and does not simulate any product design situation, unless one is designing a product to fail at a specific temperature, under load; therefore, this test should be used for reference only.

The test consists of placing a specimen edgewise as a beam over two support points 4 in. (10.2 cm) apart (Figure 10-17). This apparatus is placed in a heated liquid bath capable of maintaining the estimated HDT for the specific plastic being tested. The temperature of this bath begins at room temperature and is increased at the rate of 3.6° F (2° C) per minute. A load is placed against the edge of the specimen and the temperature at which the specimen deflects to a total of 0.010 in. (0.025 cm) is recorded as the heat deflection temperature.

Vicat Softening Temperature—ASTM D-1525 (ISO 306:1987)

The Vicat softening temperature test is similar to the HDT test, but the specimen is not placed on edge or on support beams; it is placed flat at the base of the apparatus, which is then immersed in a suitable heated liquid bath (Figure 10-18). A needle probe is placed against the surface of the plastic and a specific load is applied to the probe. The temperature of the bath is slowly increased until the needle penetrates the plastic specimen to a total depth of 0.040 in. (0.1 cm). This is recorded as the Vicat softening temperature.

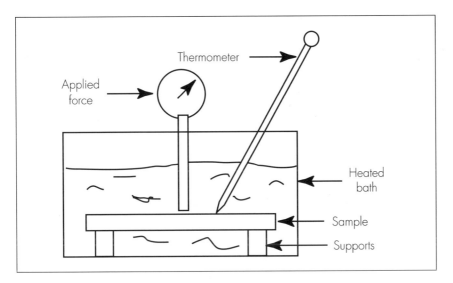

Figure 10-17. Heat deflection temperature testing.

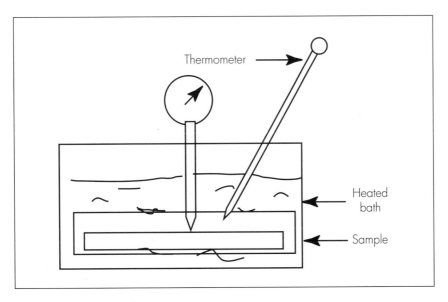

Figure 10-18. Vicat softening temperature testing.

Flammability—ASTM UL-94 (ISO UL-94)

The accepted test for flammability is performed under the guidelines presented by Underwriters Laboratories (UL). These tests are performed on

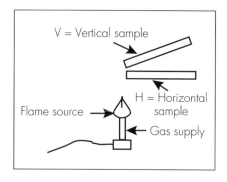

Figure 10-19. Flammability testing.

specimens placed either horizontally (H) or vertically (V) with respect to a flame source, depending on the plastic being tested (Figure 10-19).

The measurements that must be recorded include the condition of the plastic when an ignition source is applied; then the condition at the point the ignition source is removed; then again as the ignition source is reapplied. Any smoking, dripping, or other problem is recorded, and the speed and distance of any flame travel is calculated. The test is run numerous times and numerical values are stated for the various conditions noted. These are added together and the UL rating is specified as a result of the total value.

It must be noted that this test is *not* to be performed on a product, but only on a specially molded sample of the specific material being tested. Also, the rating must state a thickness at which the rating applies. This is normally 1/16 in. (0.16 cm), but other thicknesses are acceptable. Generally speaking, the lower the thickness chosen, the more flame-retardant the plastic.

Limiting Oxygen Index (LOI)—ASTM D-2863 (ISO 4589:1984)

This test is being used, in many cases, to replace the UL 94 flammability test because it appears to be applicable to molded products, unlike the UL test which requires specially molded specimens. The LOI test is used to measure the minimum amount of oxygen that will support flaming combustion of a plastic product.

The test consists of mounting a specimen vertically in a tube (Figure 10-20). A flow of air is passed

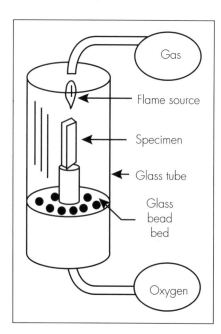

Figure 10-20. LOI testing.

through the tube containing a specific percentage of oxygen. The specimen is ignited with a flame source, then the source is removed and recording is begun. The oxygen level is adjusted upward or downward to determine the minimum level that will sustain burning of the specimen for a specific amount of time. This level is stated as the percentage of oxygen contained in the airstream.

FAILURE ANALYSIS

While failure analysis (FA) largely concerns finished products that may have failed during use, it also can be considered a method of analyzing product design and stability to determine causes of defective (but not failed) parts, as well as causes of defective processes related to molding and finishing that part. Failure analysis can be used to predict what might happen if certain actions are taken regarding design changes, tool changes, material changes, or process changes. Failure analysis differs from troubleshooting in that FA is generally conducted after the product is molded; troubleshooting is usually performed during the molding process.

Overview

Failure analysis is a concept rather than a method. It may require sophisticated instruments and tools in some cases, but in many cases visual observation is all that is necessary, such as analysis of surface cracks resulting from normal use of a flat plaque. Visual observation may detect that one corner of the plaque was being held up in the mold during ejection. This would cause a stress condition in that corner which might not be visible during normal inspection procedures but would be emphasized if the part was ever physically struck on that corner, releasing the stress and resulting in cracks.

Failure analysis can be thought of as a form of reverse engineering. A failed part is inspected, an initial determination is made as to what analysis equipment, if any, is required, and samples are gathered or created. In addition, as much data as possible is recorded regarding conditions existing at the time of the failure. This may require a separate investigation.

Stress

Most failures of plastic products can be traced to stress. In Chapter 4, we defined stress as *a resistance to deformation from an applied force*. All this means is that the plastic molecules (either molten or solidified) are trying to travel one way and something is trying to force them to travel a different way. This produces stress, the five most common forms of which are tension, compression, bending, twisting, and shear.

These forms of stress can be created by such situations as:

- Moisture in the material,
- Improper temperature profile in the heating cylinder,
- Hot and cold spots in the mold (more than 10° F [(5.6° C]) difference between any two points.
- One-half of the mold hotter than the other,
- Gating into a thin section rather than a thick section of the part,
- Improper injection pressure profile,
- Inconsistent cycles,
- Material degradation,
- Excessive regrind usage,
- Physical obstructions in the mold such as burrs on metal edges,
- Improper runner design.

It should be obvious that if so many things can cause stress, stress is probably going to occur on a regular basis. And, when stress is present, there will be product failures. Stress conditions must be minimized as much as possible, and methods for doing so are discussed in Chapter 4. But, with all of the awareness and minimization efforts, failures due to stress and other situations may still occur. Several methods and types of equipment are available for determining the causes of those failures.

Differential Scanning Calorimeter

The Differential Scanning Calorimeter (DSC) (Figure 10-21) compares the amount of energy (in calories) required to establish a zero temperature difference between a substance (plastic) and a reference specimen. The DSC generates a curve that can be plotted and analyzed. As with other testing, sample preparation is the key to successful DSC testing. The sample size is normally below 20 mg, and samples can be prepared either from raw material (regrind or virgin), or from the molded product.

Stress

If stress is present in a molded part, the stress will be defined as a *spike* on the DSC curve. Figure 10-22a shows this spike as it appears on the curve for a crystalline material. It will also appear on the curve for an amorphous material, but will not be as evident.

Moisture

If moisture is present in a raw material, either virgin or regrind, it, too, will show up on a DSC curve as a spike, but the spike will be located at the 212° F (100° C) mark, where the moisture turns to steam (Figure 10-22b).

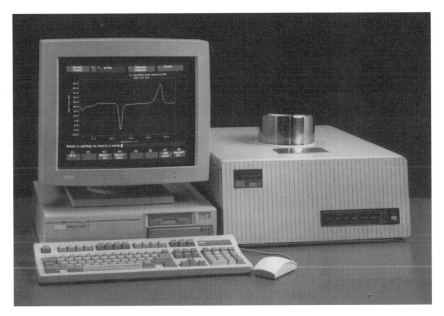

Figure 10-21. Differential Scanning Calorimeter. (Courtesy Perkin-Elmer Corp.)

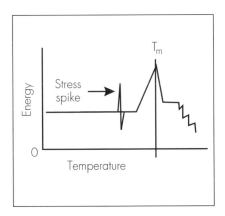

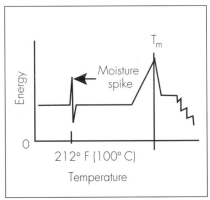

Figure 10-22a. Stress spike on DSC curve for a crystalline plastic.

Figure 10-22b. Moisture spike on DSC curve for an amorphous plastic.

T_m and T_g Points

The DSC curve displays the melting temperature of crystalline materials and the glass transition temperature of amorphous materials. See Figures 10-16a and 10-16b for examples of these DSC curves.

Regrind Percentage

Excessive regrind can cause degradation and brittleness in the product. The product can be analyzed by running a DSC curve that will display two peaks (crystalline materials) or two transition points (amorphous).

Figure 10-23 shows the two DSC peaks. The one on the right is for virgin and the background one on the left is for regrind. There are two peaks because regrind melts at a lower temperature than virgin material. The DSC will calculate the area beneath the two peaks. Regrind percentage is calculated by establishing what percentage of the total area is composed of the regrind area.

Crystallinity

The crystal structure of a crystalline material breaks down when exposed to the heat of the injection barrel. For the molded product to attain maximum structural properties, this crystal structure must be allowed to re-form as the material cools and solidifies. If the material cools too quickly, only a portion of the crystals is allowed to re-form, and structural integrity suffers. The usual amount of recrystallizing that is acceptable is 85 percent. The DSC will determine crystallinity percentage through a comparison of two samples: a known, fully crystallized specimen and a specimen taken from the part being tested.

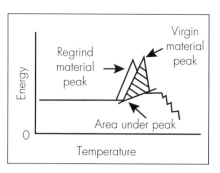

Figure 10-23. Determining regrind (crystalline material).

To calculate crystallinity, the area beneath the peak of the sample being tested is compared to the area beneath the peak of the known sample (Figure 10-24). The ratio that develops converts to the percentage of crystallinity that was attained when the tested sample was originally molded. This same method is used to determine the degree of cure of thermoset materials.

Calculating Glass Content

Furnace Method

A relatively simple test, the furnace method of determining glass content is performed by first weighing the molded sample, then placing the sample in a high-temperature muffle furnace (Figure 10-25), capable of maintain-

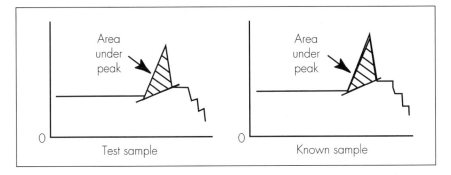

Figure 10-24. Crystallinity curves.

ing approximately 1000° F (538° C) for extended periods of time. After 30 minutes to 2 hours, the sample is removed, allowed to cool to room temperature, and weighed again. The difference between the first and second measurements is the percentage of resin, binder, and fillers that were burned away. The remaining material represents the percentage of glass reinforcement that is present. This test can also be performed using a hand-held gas torch in place of the furnace.

Thermogravimetric Analysis (TGA) Method

Thermogravimetric analysis is a process that can be used to determine filler content, resin content, and content of other components of a molded

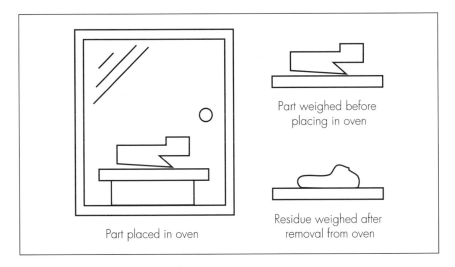

Figure 10-25. Samples in furnace.

product or raw material. The process consists of placing a small specimen in a chamber that is part of the TGA apparatus (Figure 10-26a). This chamber continuously weighs the sample and creates a curve displaying that weight. The chamber is slowly heated to approximately 1000° F (538° C).

Figure 10-26a. TGA chamber. (Courtesy Perkin-Elmer Corp.)

As the sample heats, the components in the plastic are burned. The curve shows a peak at each point when this happens, until there is nothing left but ash. This ash is the reinforcement material (fiberglass) that was used in the plastic.

The curve that is generated (Figure 10-26b) shows the weight of residue left in the chamber at each peak, so a percentage can be calculated from that data.

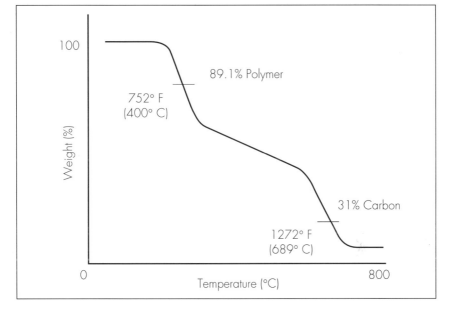

100

89.1% Polymer

752° F
(400° C)

Weight (%)

31% Carbon

1272° F
(689° C)

0

Temperature (°C)

800

Figure 10-26b. TGA curve. (Courtesy Perkin-Elmer Corp.)

SUMMARY

Proper conditioning of the samples is the most important item to remember for performing tests on plastic parts or materials.

The five most common electrical tests are dielectric strength, dielectric constant, volume resistivity, surface resistivity, and arc resistance. The five most common physical tests are shrinkage rate, density, water absorption, moisture content, and melt flow index. The five most common mechanical tests are tensile strength, compressive strength, flexural strength, creep, and impact resistance. The five most common thermal tests are melting point, heat deflection temperature, Vicat softening temperature, flammability, and limiting oxygen index.

Failure analysis differs from troubleshooting in that FA is usually performed after the product is molded, while troubleshooting is usually performed during the molding process.

Stress is the number one cause of product failure. Stress is a resistance to deformation from an applied force and, as such, is the result of molecules being forced to travel in directions and by methods that oppose their natural tendencies. A Differential Scanning Calorimeter can be used to indicate the presence of stress in a molded part. The DSC can also be used to test for the presence of moisture in raw materials, the percentage of regrind present in a molded part or raw material mix, and the degree of crystallinity present in a molded product.

The percentage of glass reinforcement available in a molded part or raw material can be determined by either the furnace method or the thermogravimetric analysis method. Both methods consist of burning away all the organic material from the plastic and leaving the glass.

QUESTIONS

1. What are the two primary sources of information regarding test procedures for the plastics industry?
2. What will improper sample preparation cause?
3. Name three of the five common electrical tests.
4. Name three of the five common physical tests.
5. Name three of the five common mechanical tests.
6. Name three of the five common thermal tests.
7. What is the principal difference between failure analysis and troubleshooting.
8. What is the definition of *stress* as it is used in this book?
9. What is the full name of the test equipment known as DSC?
10. In your own words, how is regrind percentage determined with the DSC?
11. How is crystallinity percentage determined with the DSC?
12. Name the two common methods for determining percentage of glass content.

Troubleshooting

11

OVERVIEW

Too often, a plastics technician, engineer, or operator will be presented with a molding problem and will start turning dials, flipping switches, and adjusting timers without understanding what is being done or knowing what results to expect. This is common and results from an instinct to do something (anything) because a quick fix is wanted, although not always possible. It doesn't have to be that way. The troubleshooter can objectively analyze a molding defect and eventually come up with a potential solution. He or she should, of course, put the potential solution to the test and follow it up by another analysis. Either the potential solution worked or it didn't, in which case another solution should be developed. But each solution should be determined independently and rationally. There should be no guesswork, and assistance from outside sources should be sought and welcomed.

One popular source of troubleshooting assistance is the material suppliers, who can provide detailed guidesheets about what to do if certain defects are encountered. Though the guidesheets do not provide causes for every problem, they are well researched, and a troubleshooter may eventually find the answer to a specific problem.

It is better to use a two-edged approach to troubleshooting that consists of using the material suppliers' guides and just plain common sense. That's the approach taken in the following segment.

WHAT CAUSES DEFECTS?

A study that took place over a 30-year span (1963 to 1993) by Texas Plastic Technologies* analyzed the root causes of the most common injection-molding defects. The defects studied were process-related and did not include those resulting from poor basic product design. The study found that the defects could be traced to problems with one or more of the following four items: the molding machine, the mold, the plastic material,

---

*Texas Plastic Technologies, 605 Ridgewood Road West, Georgetown, Texas 78628.

and the molding machine operator. Of particular interest was the percentage that each of these items contributed toward the defects. Figure 11-1 shows the breakdown.

Many of us in the industry believe that the most frequent cause of defects is the material, with the operator coming in a close second. But as Figure 11-1 shows, the most frequent cause of defects is actually the molding machine. Thus, when troubleshooting, the first place to look for a solution to a defect problem is the machine, because the answer will be there 6 out of 10 times.

A troubleshooter must be able to approach a problem with an objective mind. What solved a problem one day may not solve the same problem another day. Because of the large number of parameters and the variability of these parameters, and the way they all interact, many solutions may exist for a single problem. Likewise, many problems may be fixed by using a single solution. So, the troubleshooter must think through

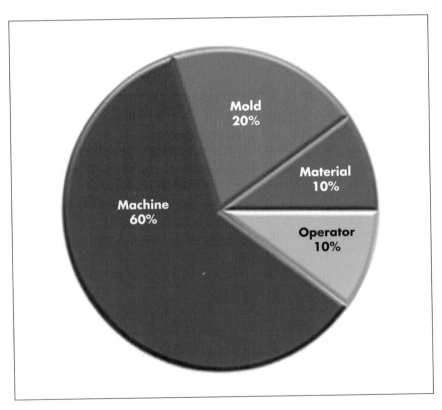

Figure 11-1. Distribution of defect causes.

the problem and make sure the proper solution is chosen. This is done by applying objectivity, simple analysis, and common sense.

The first step is to visualize the way a process should be running. Most troubleshooting is actually performed after a specific job has been running successfully for an extended period of time. There has been an initial setup and debugging process, and the mold has been accepted for production. Then, after running successfully, parts begin to be molded with defects. This is when the troubleshooter is brought into the picture. This is also when common sense and objectivity must be brought into play.

Visualizing what happens to the plastic as it travels from the hopper through the heating cylinder and through the flow path to the cavity image, you can determine what may have changed to cause defects. A heater band could be burned out, or an injection pressure valve spring may be weak, or cooling water lines may have become blocked. Any of these problems will cause specific things to happen. A thorough understanding of the molding process will help determine the cause.

On the following pages are listed 24 of the most common molding defects (along with the causes related to machine, mold, material, and operator) and the most popular remedies. Though not all-inclusive, the list contains the major causes and remedies.

DEFECTS AND REMEDIES

Black Specks or Streaks

Machine

Excessive residence time in barrel. Between 20 and 80 percent of the barrel capacity should be injected each cycle. If the plastic stays in the barrel longer than normal, it will begin to degrade. This degradation results in carbonized plastic, which appears as small black clusters, as shown in Figure 11-2. These can be carried through the melt stream and show up as spots or streaks in the molded part, visible on the surface of an opaque part and throughout a transparent part. The solution is to place the mold in a properly sized machine.

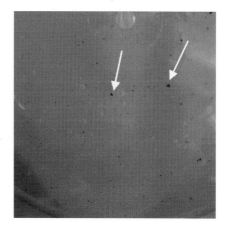

Figure 11-2. Specks are caused by overheating or contamination of the material.

Mold

Sprue bushing cracked, nicked, or not seating properly. Any of these conditions will cause plastic to hang up in the crack, nick, or offset seat of the bushing. The material can overheat due to excessive residence time at that location, and this can cause degradation or carbonizing. Eventually the hung-up resin breaks loose and enters the melt stream and flow path. The remedy is to replace cracked or nicked bushings, and use a blueing agent to check that the bushing is centrally seated against the nozzle tip. Also, check that the nozzle tip opening has an equal or smaller diameter than the sprue bushing to ensure a proper seal.

Material

Contaminated raw material. Such contamination can be the result of dirty regrind, mixed regrind, improperly cleaned hoppers or grinders, open or uncovered material containers, and even poor-quality virgin material from the manufacturer. The remedies include dealing with only high-quality suppliers, using good housekeeping practices, and properly training material handling personnel.

Operator

Inconsistent cycles. The operator may inadvertently be causing delayed or inconsistent cycles. This will result in either excessive residence time of the material in the heating cylinder, or overcompensating heater bands. Both conditions will result in degraded material, especially with heat-sensitive plastics. One remedy is to place the machine in automatic mode, with the operator serving as monitor to stop the press if an emergency develops. The operator should be trained to be aware of the importance of consistent cycles, whether or not the machine can run automatically.

Blisters

Machine

Back pressure too low. As the material is heated and augured through the heating cylinder, air becomes trapped within the melt. One of the uses of back pressure is to force this air out before it gets injected into the mold cavity image. Back pressure should be set at 50 psi (345 kPa) and increased in increments of 10 psi (69 kPa) until the ideal setting is reached.

Mold

Mold temperature too low. As a material is injected into a mold, it starts to cool immediately and a skin begins to form on the surface of the part. If this skin forms too quickly, any air that is mixed into the material will not be allowed to escape through the surface as intended, causing a blister effect (Figure 11-3). A mold that is too cool will cause the skin to form too soon. Increasing the temperature of the mold will help allow trapped air to escape by delaying the hardening of that skin.

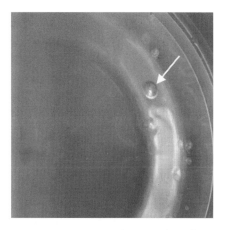

Figure 11-3. Trapped air resulting from process deficiencies causes blistering.

Material

Use of regrind that is too coarse. This practice increases the amount of air that gets trapped in the melt because the coarse, uneven particles of regrind create pockets of air between them and the smaller, consistently sized particles of base material. One remedy is to use a finer-gage screen in the regrinder. Another remedy is to limit the amount of regrind that is used to less than 5 percent. Or you can increase the amount of back pressure on the injection screw, assuming the base material is not too heat-sensitive. Another solution, if others fail, is to use only virgin material. In fact, sometimes this can be done to start the run and regrind can be "salted in" as the run progresses.

Operator

Early gate opening. There is a slight possibility that blisters will form if the operator were to open the gate too soon, thus not allowing the part to cool (solidify) in the mold. This would have to be precisely timed, however, as the part probably would warp, twist, or otherwise deform drastically before blisters would form.

Blush

Machine

Injection speed too fast. The speed and pressure of the melt as it enters the mold determine both density and consistency of melt in packing the mold. If the fill is too fast, the material tends to slip over the surface, especially at the gate area (which may cause gate burn), and the material at the slipped surface will skin over before the rest of the material solidifies. This area will not faithfully reproduce the mold steel surface as does the material in the other areas of the part because it has not been packed as tightly against it. Thus the underpacked area has a duller finish (Figure 11-4). The injection speed must be adjusted (decreased) until the optimum has been reached, which may require adjustments in barrel or mold heats as well.

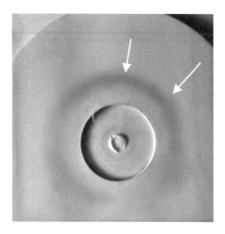

Figure 11-4. Blush can result from density, temperature, pressure, and cycling shortfalls.

Mold

Mold temperature too cold. If the mold is too cold, the flow of the molten material is hindered and the material solidifies before it fills and packs the mold. Blush (dull finish) will appear in the last area to be packed, usually the gate. Increasing the mold temperature allows the material to flow farther and pack properly.

Material

Excessive moisture. Excessive moisture in a melt may accumulate in the gate area because injection pressure tends to force trapped moisture out of areas that have been packed and push it into unpacked areas. The gate is the last area to pack, so it is the last place moisture may collect. This causes the area to look dull. The blushing may be accompanied by splay, or silver streaking.

Operator

Inconsistent cycling. Inconsistent cycling of the molding press by the operator may cause the material to overheat. If this happens, it is possible that the injection fill rate will increase in random cycles. The effect of this increase is explained under "Machine" above. Operating the machine on automatic cycle mode helps to ensure consistent cycles.

Bowing

Machine

Clamp opens too quickly. To increase the number of cycles produced in an hour, molders may sometimes increase the speed at which the clamp opens the mold at the end of the molding cycle. If this is done at the very instant of the mold-open portion of the cycle, there will be a tendency for the part to hang up on the injection half (dead half) of the mold. As the mold continues to open, the part will snap back onto the clamp half (live half) of the mold, and the result may be a bowed part from this distortion (Figure 11-5). The solution is to make sure the first 1/4 in. (0.64 cm) of mold opening is set at a slow speed. The balance of the opening cycle can then be set at a much faster speed.

Mold

Temperature too low. Some materials, such as certain polyesters, require mold temperatures that are above the boiling point of water (212° F [100° C]) in order to achieve maximum physical properties of the

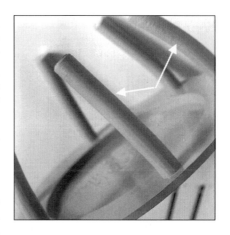

Figure 11-5. Bowing defects can be traced to the machine, the mold, or the operator.

materials. Parts that are molded at too low a mold temperature are not physically strong enough to overcome the tendencies to bend when the mold opens. The result might be bowing. Raise the mold temperature to that recommended by the material supplier for the specific resin being molded. This may require the use of an oil heater or electrical heaters placed in the mold.

Material

There is no indication that the material itself is the cause of bowing—except as related to shrinkage characteristics, and generally these are controlled through other means.

Operator

Improper handling. The operator may improperly handle molded parts after ejection from the mold. If parts are packed for shipment too soon after molding, the heat that they retain may not be allowed to dissipate properly and they could take a bowed set. Also, relief operators may not handle the parts the same way the main operator does. Proper instruction is required to remedy that situation, and the packaging process should be analyzed and corrected as appropriate.

Brittleness

Machine

Improper screw design. A screw with too low a compression ratio for the material being molded will not properly melt and mix the material. This

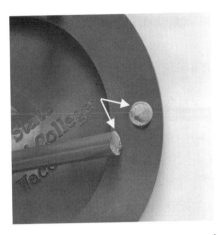

Figure 11-6. Melt, mix, moisture, and molding inconsistencies cause brittleness.

results in weak bonds between the individual molecules in the material and the part exhibits brittleness (Figure 11-6). Use of an injection screw with a higher compression ratio will help solve this problem. The material supplier is the best source to contact for the proper screw design for specific materials.

Mold

Condensation. Although it does not occur with any regularity, condensation in the mold cannot be ruled out as a possible source of moisture, which in turn may cause brittleness in molded parts. This condensation will be especially prevalent in molds that are operated under humid conditions. Cooling water in the mold may be the source of such condensation. One remedy is to use insulation panels between the mold and the press, as well as on all the outside surfaces of the mold. Another is to raise the mold temperature slightly to reduce the tendency to form condensation. A small fan blowing around the mold may be of some benefit, but it should *not* blow directly on the molding surfaces of the mold.

Material

Excessive moisture. *All* materials need a small amount of moisture in order to be processed properly, but this is usually in the area of 1/10 of 1 percent. Some materials such as nylon and acrylonitrile-butadiene-styrene (ABS) are hygroscopic by nature and readily absorb moisture from the atmosphere, even after initial drying. These are difficult materials to keep dry. Moisture causes brittleness because the water droplets turn to steam when heated in the injection unit and this steam explodes through the melt stream, causing voided areas. These voided areas are not properly bonded and easily break apart when they are subjected to any mechanical forces after molding.

Some materials (especially hygroscopics) may require conditioning after molding to put back the moisture that was removed during the mold process. Nylons, for example, normally must be conditioned by either annealing in 300° F (149° C) glycerin for 4 hours, or being placed for 4 days in sealed bags filled with water. Without this conditioning, the plastic will be brittle as the result of *proper* drying procedures used to mold the plastic.

Operator

Inconsistent cycles. An operator who is controlling the cycle may cause brittleness if the machine is not kept cycling consistently from shot to shot because the material will tend to degrade in the heating cylinder. Degraded material causes weak molecular bonding, which results in brittle parts.

Bubbles (Voids)

Machine

Injection temperature too high. High injection temperatures can cause the molten material to be too fluid. This may result in the material being so turbulent that air and gases become trapped in the melt stream. The trapped gases show up as voids in the molded part, as shown in Figure 11-7. Reducing the injection temperature allows the material to stiffen, permitting the trapped gases to escape from the melt stream. *Caution:* Apparent voids may sometimes turn out to be unmelted particles. If that is the case, reducing the temperature will only make the condition worse; increasing the temperature will help melt the particles.

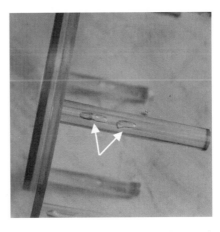

Figure 11-7. Bubbles form as the result of air or gases trapped in the melt stream.

Mold

Section thickness too great. When a plastic part consists of varied wall thicknesses (instead of one steady thickness), the thicker walls will cool (and solidify) last. There will be a pressure loss in those thick areas as they continue to cool after the thinner areas have solidified. The plastic will pull away toward the solid section and cause a void in the thick section. When the void is on the surface of a part, it appears as a sink mark. When it is below the surface, it appears as a bubble. The best solution (although expensive) is to use metal core-outs to thin the thicker wall. Or, if possible, change the wall thickness so that the thicker section is no more than 25 percent thicker than the thin section. This will minimize the void.

Material

Excessive moisture. Excessive moisture can get trapped in the resin as the molding process progresses and show up as bubbles in the molded part. The moisture actually turns to steam during the heating process and cannot escape from the material, so it forms a gas pocket that becomes a void. The obvious solution is to properly dry the material before molding.

Operator

Inconsistent cycles. This may cause the temperature controllers for the heating barrel to overrun, thus making the material too fluid. As a result, the material may be injected at too high a speed which may cause gases to be trapped. These will then show up as pockets (voids). Ensure consistent cycles by running the molding machine in an automatic mode whenever possible. If this is not possible, instruct the operators so they know the results of running inconsistent cycles.

Burn Marks

Machine

Excessive injection speed or pressure. If injection pressure is too high, the resin is forced into the mold so fast that any air trapped in the runner system or mold cavities is not allowed time to be pushed out ahead of the resin flow. Then this trapped air becomes compressed and its temperature rises sharply. The hot air ignites the surrounding plastic resin, which burns until the air is consumed, leaving a blemish like that shown in Figure 11-8. Reducing the injection speed and pressure will allow enough time for the gas or trapped air to escape through normal venting methods.

Figure 11-8. Heat buildup during molding can ignite material and create burn marks.

Mold

Improper venting. Venting systems are placed in molds to exhaust any gases or trapped air that might be present. If the vents are not deep enough or wide enough, or if there are not enough vents, the air is compressed before it is all exhausted and then it ignites and burns the plastic as described under "Machine" above. Vents must be a minimum of 1/8 in. (0.3 cm) wide. The vent land should not be more than 1/8 in. long. Blind areas, such as the bottom of holes, should have vents machined on the side of ejector pins that are placed there. There should be enough vents on the parting line to equal 30 percent of the distance of the parting line perimeter. Thus a 10-in.-long (25.4-cm-) parting line perimeter would have 12 vents, each 1/4 in. (0.64 cm) wide (3 in. [7.62 cm] total).

Material

Excessive regrind use. The use of regrind may have to be limited, especially with heat-sensitive materials such as polyvinyl chloride (PVC). Regrind material tends to absorb heat in the injection barrel at a slower rate than virgin, because of the irregular surfaces and larger size of the regrind particles. This results in a longer heating cycle which causes the virgin pellets to overheat and degrade. The degradation takes the form of burned particles which are transported through the melt stream into the cavity. Limit regrind use to no more than 5 or 10 percent. If the volume of shot size is small (less than 20 percent of barrel volume), it may require no regrind at all. A possibility is to start with all virgin and slowly build up regrind use by salting in regrind at 2-percent increments until burning occurs. Then drop back 2 percent and use the resultant ratio for future molding.

Operator

Inconsistent cycles. Erratic cycles cause the barrel heating system to heat in erratic steps, resulting in hot spots in the barrel. In these areas, the material is overheated and degraded. Again, the degradation takes the form of burned particles which are transported through the melt stream and into the cavity. If possible, run in automatic mode. If not, at least instruct all operators on the importance of running consistent cycles, demonstrating the burning effect.

Clear Spots

Machine

Barrel temperature too low. Low barrel temperatures result in an improper blending of molecules due to unmelted particles. These particles travel through the melt stream and enter the cavity. In transparent parts, they show up as clear spots, but even in opaque parts, they may show if near

Figure 11-9. Insufficient melting of the material can cause clear spots to appear.

the surface, as indicated in Figure 11-9. Cutting open an opaque part will reveal the spots as voids. Increasing barrel temperature will reduce the tendency for unmelted particles. Increase in 10° F (5.6° C) increments and allow the temperatures to stabilize (10 cycles) before increasing again.

Mold

Water leaks. There is a possibility that the mold has developed cracks that may allow water to seep into the cavity from cooling lines in the mold. If this happens, the water drops may appear as clear spots in transparent parts. Check the molding surfaces of the mold to see if any moisture is evident. A pressure check of closed water lines will determine if cracks (thereby leaks) are present. If not possible to weld these cracked areas, it may be possible to use tubing inserted in the lines to stop leaking. The condition causing the cracks should be rectified and the mold base should be reinforced if it is considered usable.

Material

Excessive regrind. Because regrind material absorbs heat slower than virgin (owing to irregular particle sizes), there is a tendency for the regrind not to melt as well as the virgin under normal heat settings. Increasing the barrel temperature slightly may be enough to accommodate the regrind if the material being molded is not too heat-sensitive. But be careful that the virgin material does not become overheated and degraded.

Operator

Inconsistent cycles. Erratic cycling of the machine can result in erratic heating in the barrel, causing hot spots and cold spots. Material from the cold spots may continue through the melt stream without being properly heated and will show up as clear spots in transparent parts. Training the operators and explaining the importance of consistent cycles should be accompanied by examples of defective parts run during inconsistent cycling.

Cloudy Appearance

Machine

Barrel temperature too low. A cloudy appearance, especially in a transparent part, must not be confused with blush. If a true cloudy appearance is evident, like that in Figure 11-10, it is normally due to a group of improperly melted particles. These are not blended with the main melt and tend to isolate themselves in a group pattern. Increasing the barrel temperature reduces the likelihood of unmelted particles making it into the melt stream, but make sure this increase does not degrade the virgin resin particles.

Mold

Uneven packing of cavity. Uneven packing can normally be traced to improper gate or runner sizing or location. The material enters the cavity at the wrong spot, preventing it

Figure 11-10. Cloudiness is caused, again, by improper melting of material.

from being packed against the mold steel in all areas. The material solidifies without replicating the mold finish, and this appears as a cloudy area. Also, if one area of the molding surface was not polished as well as the other areas, material in that area would appear cloudy. It is important that the mold is properly polished. Then use a computer program to determine the proper gate size, number, and location for a specific product design. In the absence of a computer, it is possible to determine proper parameters through trial and error, but it is very time-consuming and expensive. A design consultant may be a good investment.

Material

Excessive moisture. Moisture turns to steam as it progresses through the heated barrel of the molding machine. As it enters the cavity, it literally explodes against the molding surfaces. This usually takes the appearance of splay or silver streaks, but sometimes appears as clouds. Make sure the material to be molded has been properly processed for drying according to the properties of the material.

Operator

Inconsistent cycles. Inconsistent cycling can cause temperature controllers to become erratic and create hot and cold spots in the heating barrel of the molding machine. The cold spots result in unmelted particles that travel through the barrel and into the cavity. Groups of these particles may form in specific areas and appear as a cloud on the surface. Maintaining consistent cycles will minimize the erratic heating. Inform operators of the importance of consistent cycles, and run in automatic mode if at all possible to eliminate the operator's influence.

Contamination

Figure 11-11 shows dramatically the effect that contamination can have on a part or product. Contamination in injection molding can originate in any of a number of areas.

Figure 11-11. Good maintenance and housekeeping can prevent material contamination.

Machine

Oil leaks. It is common for molders to allow small oil leaks to become big ones before they consider fixing them. Leaking oil has a tendency to find its way into some unbelievable places such as the feed throat of the injection barrel during material changes. Also, when a machine is lubricated, the greasing is usually overdone and grease drips end up on mold surfaces and machine areas, finding their way into the plastic material. Eliminate oil leaks and clean up grease drips. By fixing oil leaks, wiping up grease drips, and cleaning up chemical spills, these sources of contamination will be greatly minimized.

Mold

Excessive lubrication. Molds with cams, slides, lifters, and other mechanical actions need periodic lubrication. The tendency to overdo this allows the lubricant to find its way to the molding surfaces and enter the molded part. Also, excessive use of mold release causes contamination problems. Use the proper lubricant for specific mold components and use only the amount necessary. Optimize the use of mold release if it is needed at all. It seems to be human nature to think that if a little works, a lot will work better, but that is not the case with mold releases.

Material

Improper regrind. Regrind has been found to contain any number of contaminants such as residues from food containers, soft drink spills, dust and dirt particles, and other plastic materials. This is usually due to poor housekeeping and/or material handling procedures. This type of contamination can be greatly reduced by proper instruction of personnel, highly visible labeling of regrind containers, proper labeling of trash containers to differentiate them from material containers, tight fitting covers for regrind (and any plastic) material containers, proper cleaning of regrind machines, and care applied during material changeovers.

Operator

Poor housekeeping. An operator can cause contamination through actions such as snacking at the molding machine station. Potato chip salt and soft drink spills are common sources of material contamination. Dust from sweeping can enter the hopper if it is not covered. And, in rare cases, operators have been known to force a break by intentionally throwing trash in the hopper. Instruct operators on the importance of maintaining good housekeeping practices and hold supervisors to the same standards.

Cracking

Machine

Molded-in stresses. Stresses can be molded into a product through the molding machine by excessive packing or too fast a filling rate. The plastic is injected and held against the restraining surfaces of the mold cavities. When the part is ejected, the cooling process continues (for up to 30 days) and the highly pressured plastic may begin to relieve. If the skin of the molded part is not yet solid enough, it will split open in the form of cracks similar to those shown in Figure 11-12. Reduce the injection pressure and speed to the lowest numbers that will successfully mold the part. This reduces the tendency to mold in stress.

Mold

Insufficient draft or polish. Draft angles should be an absolute minimum of 1° per side to facilitate easy removal of the part from the mold. Ejector pressure may cause cracked parts if less than that is used. Also, rough cavity surfaces (and other undercuts) cause a drag on the part as it ejects.

Figure 11-12. Cracking can be caused by the machine, the mold, the material, or the operator.

This may cause cracking if the ejection pressure is increased to push the part over this rough surface. Make sure cavity surfaces have a high polish when the mold is built and that they are repolished as the need arises.

Material

Excessive moisture. Because moisture turns to steam as it travels through the heating barrel, it creates a condition in which material particles do not properly bond in the area where moisture is present. This improper bonding causes weak areas in the molded part, and these may crack when parts are handled. Again, it is important to ensure proper drying of the material prior to molding to minimize the tendency to crack due to moisture content.

Operator

Inconsistent cycles. Erratic cycling will cause hot spots in the heating barrel. The material in these areas may degrade due to overheating. This degraded material will not properly bond with surrounding material particles and will cause weak areas in the molded product that may develop into cracks. Inform (and show) operators that inconsistent cycles will cause defects such as cracking and encourage them to use automatic cycling whenever possible to eliminate the influence of operator inconsistencies.

Crazing

Because crazing (Figure 11-13) is simply a very fine network of cracks, the same causes and remedies apply that are mentioned under "Cracking," in the previous example.

Delamination

Machine

Injection speed too slow. Injection speed determines how fast the material is injected into the mold. If it is too slow, the material tends to cool off and solidify before the mold is filled. Because the material fills the cavity in a tongue-on-tongue fashion, a layer may begin to solidify before the next layer bonds to it. This results in a separation after the part is ejected from the mold and manifests itself as delamination, creating a scrap part like the one shown in Figure 11-14. Increase injection speed in small increments (2 percent of the total speed) until delamination is eliminated. If flashing or burning occurs, the limits have been reached and another source of the defect should be investigated.

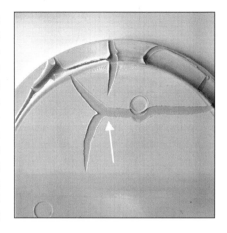

Figure 11-13. The causes of crazing (network cracking) are similar to those of cracking.

Mold

Mold temperature too low. If the mold temperature is too low, the incoming layers of molten material may cool off too soon and not bond to each other. On ejection, these unbonded layers separate, causing delamination. Increase the mold temperature in 10° F (5.6° C) increments until the delamination is eliminated. Then increase by an additional 10° F step to compensate for thermal fluctuation in the mold.

Material

Foreign materials or additives. If a pigment is being used to color the resin, it may not be compatible, such as where soap is used in the manu-

Figure 11-14. Insufficient bonding is the principal cause of delamination.

facturing of the pigment. If a color concentrate is used to color the resin, it must be of a material compatible with the base resin. And, if accidental mixing of two incompatible resins has occurred, they will not bond. In all these cases, nonbonding of the materials used will result in delamination of the molded layers of the finished product. Check with suppliers of any additives to make sure the proper grade is being used. Also, confirm that all materials are properly identified to ensure that incompatible materials are not being mixed.

Operator

Excessive mold release. If a mold release is required at all, it is necessary to limit its use. Too much mold release will cause a penetration of molded layers by the mold release itself. This will keep the layers from bonding and result in delamination. Keep mold release away from presses unless absolutely necessary, and then use only as a quick fix until the cause of the sticking can be rectified. Operators should be made aware of the problems caused by excessive mold release use.

Discoloration

Machine

Excessive barrel residence time. The amount of time that the plastic material stays in the heated barrel until it is injected into the mold is called the *residence* time. If the individual shot size is less than 20 percent of the barrel capacity, degradation of the plastic will eventually occur. This degradation will cause a darkening of the color in light-color materials and a graying effect on dark-color materials (Figure 11-15). To prevent this, optimize the shot-to-barrel ratio by moving the mold to a press that will create a shot size that is as close as possible to 50 percent. The range should be from 20 to 80 percent.

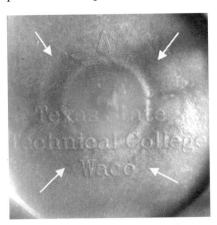

Figure 11-15. Attention to the machine, the mold, and the material will prevent discoloration.

Mold

Improper mold temperature. In general, a hot mold will cause the material to stay molten longer and allow the molecules to pack tighter. This results in a very dense part that appears darker because of that density. On the other hand, a cold mold will cause a loss in gloss because the material cools before it can be forced against the mold surface, and this will cause a less dense part that will appear lighter. Adjust the mold temperature, within the limits set for the particular plastic, to the point at which the material has the proper shade and gloss properties. Be aware that the color will change slightly after the part has been completely cooled.

Material

Contamination. Material may appear discolored if it is contaminated with any number of items including wrong regrind, fabric strands, thermally degraded material, and food particles. Also, if the entire shot of material has been exposed to excessive temperatures, it will be darkened. Proper housekeeping will minimize most of this type of discoloration, and closer control of proper molding temperatures will minimize the rest.

Operator

Inconsistent cycles. Erratic cycling will create hot spots in the heating barrel. These will cause degraded material in those local areas and this will travel through the melt stream into the cavities. Patches of this degraded material will appear discolored on the molded part. Impress on the operator the importance of maintaining consistency during molding.

Flash

Machine

Excessive injection pressure. It's possible that too much injection pressure will partially overcome the clamp pressure of the machine and cause the mold to open slightly during the injection phase. If this happens, a small amount of plastic actually seeps out of the mold. This seepage is called *flash*. An example is shown in Figure 11-16. Also, excessive pressure may force plastic into the clearance hole around ejector and core pins. This is also flash. Reducing the injection pressure minimizes flashing conditions. If the mold design allows, begin the molding process with very low pressure and slowly increase from shot to shot until the cavities are filling properly. This should be done in 50-psi (345-kPa) increments when the mold is almost filled.

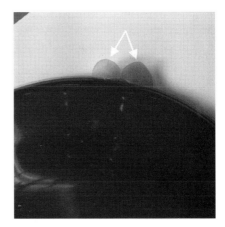

Figure 11-16. Seepage called flash can result from improper injection pressure, mold design, and material flow rates.

Mold

Inadequate mold supports. Components called *support pillars* are used in the construction of a mold to supply extra support behind the cavity retainer plates on the ejector half of the mold. They are used to keep the mold from collapsing during the injection phase of the molding cycle. If there are too few pillars, or they are not properly designed or located, the mold will tend to deflect when the injection pressure is applied. The mold will open slightly and flash will occur. Calculating the number and size of support pillars can be done with some elaborate formulas, but the following example gives some rules of thumb that are useful.

If a 12- × 15-in. (30.5- × 38.1-cm) mold base is constructed with *no* support pillars, the amount of projected part area that the mold could produce without deflection would be 14 in.² (90.3 cm²). As a general rule, the allowed projected area will increase by 100 percent for each support pillar added. If a single row of four 1.25-in.- (3.2-cm-) diameter support pillars is located along the center of the mold, the projected area allowed increases by 400 percent to 56 in.² (361.3 cm²). And, if two rows of four pillars each are added, the area increases to 112 in.² (722.6 cm²). It is best to use a few large-diameter pillars instead of many with small diameters because the smaller-diameter pillars tend to press into the support plates.

Material

Improper flow rate. Resin manufacturers supply materials in a variety of flow rates. Thin-walled parts may require easy-flow materials, while thick-walled parts may be able to use stiffer materials. The stiffer materials are usually stronger. If a fast-flowing material is used, it may creep into small crevices where thick materials could not. Flash could be the result. But even with thicker materials, if the flow rate changes to slightly thicker yet, more pressure may be required to inject the material and this could blow the mold open, also causing flash. Use a material that has the stiffest flow possible without causing nonfill or flashing the mold. This can be predetermined somewhat by consulting with a materials supplier.

Operator

Inconsistent cycles. Erratic cycling can cause the material in the barrel to overheat slightly. This will cause the material to flow easier and it may begin to flash in areas that were not flashing before. Instruct the operators on the importance of proper and consistent cycles. Use the machine's automatic mode if at all possible to minimize the operator's influence.

Flow Lines

Machine

Inadequate injection pressure. Flow lines (Figure 11-17) may be the result of improperly bonded material. If injection pressure is too low, the tongues of material that enter the cavity are not packed together to form smooth layers against the molding surface. The material actually starts to wrinkle as one layer tries to crawl over the already cooling layer outside of it. Increase the injection pressure to force the layers together quickly while they are still hot enough to bond tightly.

Figure 11-17. Improper pressure, temperature, and flow rate can result in unsightly flow lines.

Mold

Mold temperature too low. Generally, a hot mold will allow the molten plastic to flow farther before cooling off and solidifying. This results in a very dense part that minimizes the formation of flow lines. Increase the mold temperature to the point at which the material has the proper flow and packs out the cavity properly. Start by using the material supplier's recommended temperature and increase (if necessary) in increments of 10° F (5.6° C) until the flow lines disappear. Allow the mold to stabilize for 10 shots between each adjustment.

Material

Improper flow rate. A material that is too stiff (low melt index) may not flow fast enough to pack the mold before it solidifies and the flow front may not be able to squeeze out the flow lines that form. To prevent this, use a material that has the fastest flow possible without causing flashing conditions. A material supplier can make the initial recommendation.

Operator

Inconsistent cycles. Erratic cycles cause hot and cold spots to form in the heating barrel. The material in the cold spot areas may not get to the proper temperature for molding before injection. If this happens, the material will have a slower flow rate. The slow flow rate may cause the cavity to be underpacked, which can result in flow lines. Instruct the operators on the importance of maintaining consistent cycles. Show them samples of the variety of defects caused by inconsistent cycles.

Gloss (Low)

Machine

Inadequate injection pressure. If there is not enough injection pressure, the molten material is allowed to cool and solidify before the material has

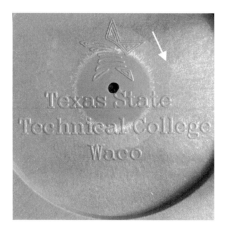

Figure 11-18. Unwanted low gloss can be prevented by maintaining proper injection pressure and flow rates.

had a chance to pack and force itself against the molding surfaces of the cavities. As a result, the material will not reproduce the finish on the mold and will simply cool down with no pressure against it. This results in a dull finish on the product similar to that shown in Figure 11-18. You can avoid this by increasing the injection pressure to force the material against the mold surface. Flow lines will be minimized and the finish and gloss of the mold will be duplicated.

Mold

Mold temperature too low. Generally, a hot mold produces a higher surface gloss on a molded product than a cold mold. This is because the particles are allowed to stay molten longer, which results in a dense product. Also, the material is able to duplicate the surface of the mold steel better because it is so dense. Increase the mold temperature to the point at which the material has the proper flow and packs the mold. This will result in higher gloss. Conversely, if too high a gloss is produced, decreasing the mold temperature will reduce the gloss.

Material

Improper flow rate. Usually, a stiffer-flowing material is favored over a less-stiff material because physical properties tend to be greater in stiffer flow materials. However, because the stiff-flow materials are harder to push, they may cool down and solidify prematurely, causing the product to be underpacked. In such a case, the surface gloss level will be much lower because the material is not packed against the mold steel. If you change to a higher-flow material grade, the material will be forced against the mold steel, duplicating the finish of that steel, and resulting in higher gloss.

Operator

Inconsistent cycles. Erratic cycling can cause cold spots in the heating barrel. The material in these areas will enter the melt stream and the cavity without being properly heated. This material will not flow properly and will not be packed in the cavity. Thus, it will not duplicate the cavity steel finish and will be dull. This may appear as spotty areas, or over the entire product, and may fluctuate from cycle to cycle. Ensure that the operator is trained in running consistent cycles; show samples of defects caused by inconsistent cycles.

Jetting

Machine

Excessive injection speed. Injection speed that is too fast will cause the molten plastic to form jet streams as it is pushed through the cavity gates instead of the more desirable wide tongue of material. These snake-shaped streams cool independently from the surrounding material and, as shown in Figure 11-19, are quite visible on the molded part surface. Reducing the injection speed will allow the plastic flow front to stay together and not form the individual streams that cause jetting patterns on the part surface.

Figure 11-19. Flow rate, mold gate location, and injection speed can be the source of the jetting effect.

Mold

Improper gate location. If material is injected directly across a flat cavity surface, it tends to slow down quickly as a result of frictional drag and cools off before the cavity is properly filled and packed out. When this happens, flow streams tend to form and the molded part surface has the familiar jetting appearance. Relocate, or redesign, the gate so that the molten plastic is directed against a metal surface instead of across a flat surface. This will cause the material to disperse and continue to flow instead of slowing down.

Material

Improper flow rate. A material that is too stiff may not flow fast enough to enter the cavity at the proper speed to maintain the desired tongue shape. It may break up into streams causing the jetting appearance. Use a material with a higher flow rate. An increase of only 2 or 3 points in melt index may be enough to eliminate the jetting defect.

Operator

Inconsistent cycles. Erratic cycling can cause cold spots in the heating barrel, and the material in those areas will flow slower than the surrounding material. These slow areas may not flow fast enough to form the desired tongue shape as the material enters the cavities through the gates. Jetting may occur as a result of the flow front breaking into separate streams of material.

Knit Lines (Weld Lines)

Machine

Barrel temperature too low. Knit lines are the result of a flow front of material being injected at an obstruction in the mold cavity. The flow front breaks up into two separate fronts and goes around the obstruction. When the two fronts meet on the other side, they try to weld back together again (knit) and form a single front once more. If the barrel temperature is too low, the material does not keep its heat long enough and, as shown in Figure 11-20, the two fronts can't make a good knit because the material has begun to solidify. By increasing the barrel temperature, you allow the melt fronts to stay molten longer and knit properly.

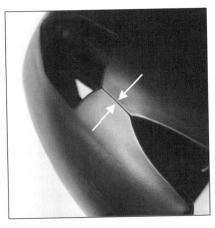

Figure 11-20. Knit lines form when material is not hot enough to reform around an obstacle.

Mold

Mold temperature too low. For the same reasons mentioned above, a low mold temperature will not allow the flow fronts to knit back together again because the material cools down too quickly. Increase the mold temperature in 10° F (5.6° C) increments until the knit line is minimized. It is impossible to eliminate the knit line, but it can be made as strong as the surrounding material by allowing a proper knitting action to take place.

Material

Improper flow rate. A stiff material will not knit together as well as a free-flowing material. This is because the stiff material moves slower and may begin to solidify before the flow fronts are knitted properly. Increasing the flow rate by 2 or 3 melt index points may be enough to attain a proper knit.

Operator

Inconsistent cycles. Erratic cycling creates cold spots in the heating barrel. The material from these areas enters the cavity gates in a sluggish pattern and the two fronts may not knit together again because the colder material has already begun to solidify. Inform the operators of the importance of maintaining consistent cycles to eliminate the cold spots in the heating barrel. If possible, utilize the machine's automatic cycling to minimize operator influence.

Nonfill (Short Shots)

Machine

Insufficient material feed. The most common cause of nonfill (Figure 11-21) is not enough material prepared in advance for injection into the mold. Increase the amount of material being fed to the mold by adjusting the return stroke of the injection screw so that more material is transferred from the hopper system with each rotation of the screw. Adjust this setting until there is between a 1/8- and 1/4-in. (0.3- and 0.6-cm) cushion at the front of the injection cylinder.

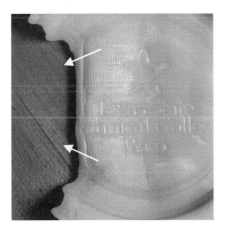

Figure 11-21. Short shots are an avoidable embarrassment caused by feed, venting, flow rate, and molding consistency problems.

Mold

Insufficient venting. Venting is used to remove trapped air from the closed mold so molten material will be able to flow into every section of the mold. If the air is not removed, it acts as a barrier to the incoming plastic and will not allow it to fill all sections of the mold. The result is nonfill. The mold should be vented even before the first shot is made. Vent the runner first, and then create enough vents on the parting line to equal 30 percent of the length of the perimeter surrounding the cavity image. An additional approach is to use a vacuum system in the mold to help pull the trapped air out before injecting material.

Material

Improper flow rate. Use of a material with too low a melt index may result in the material beginning to solidify before the entire cavity has been filled. Increasing the flow rate by 2 or 3 points may be enough to ensure the material flowing long enough to completely fill the cavity before it solidifies.

Operator

Inconsistent cycles. Erratic cycling may cause cold spots in the heating cylinder, and the material in these areas will flow at a slower rate than the surrounding material. When it enters the cavity, the slower material will solidify sooner than the rest and cause a nonfill condition. Instruct the operators on the importance of maintaining consistent cycles. Demonstrate defects caused by inconsistent cycles. Again, utilize the machine's automatic cycling if possible in order to minimize the operator's influence.

Shrinkage (Excessive)

Machine

Barrel temperature too high. If the barrel temperature is too high, the resin absorbs an excessive amount of heat. The heat causes excessive expansion of the resin molecules and increases the amount of voided area between these molecules. After injection, and on cooling, the skin of the molded product solidifies first and the remaining resin closes up the molecular voids as it cools, pulling the already solidified skin with it. The greater the amount of void volume, the greater the degree of pulling and the greater the total shrinkage. This results in defects similar to that shown in Figure 11-22. Decrease the barrel temperature to allow the resin to stay molten without creating excessive void areas. The shrinkage will return to normal. Published shrinkage data will provide the normal shrinkage for a specific material, but shrinkage rates may vary, depending on direction of flow. Material suppliers will provide range data for all directions of flow.

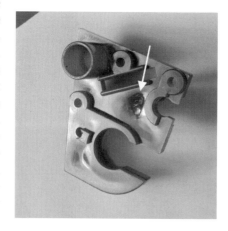

Figure 11-22. Careful control of temperature and flow rate can prevent excessive shrinkage.

Mold

Mold temperature too high. Generally a hot mold causes the material to stay molten longer, which may result in the required skin not properly forming before ejection of the product. When this occurs, the still cooling material continues to shrink because there is no normal restraining skin to hold it from shrinking too much, and the product will shrink beyond the normal dimensions. Decrease the mold temperature until the material maintains the proper flow and fills the cavity without shorting. This should be done in 10° F (5.6° C) increments, and once the desired level is reached, a single increase of 10° F should be added to compensate for fluctuation in the temperature control units.

Material

Improper flow rate. A material that is too stiff may not get fully packed into the cavity. If packing does not occur, the density of the part is too low and the part will be allowed to shrink beyond normal expectations. Raise the melt index by 2 or 3 points; the increase may be enough to allow full packing of the cavity and minimize excessive shrinkage.

Operator

Early gate opening. If the operator is speeding up the cycle by opening the gate too soon, the cooling plastic may not have had a chance to solidify enough to form a proper skin on the molded product. If this skin is not solid, the remaining plastic that cools pulls the skin with it and, with nothing restraining it, continues to shrink beyond expected rates. Instruct the operator that cycles that are too fast may cause defects in the molded product that are not even visible.

Sink Marks

Machine

Insufficient injection pressure or time. Injection pressure must be high enough to inject material into the mold and force the material to fill every part of the mold until the mold is packed solidly. When properly achieved, this packing ensures that all the resin molecules are held as closely as possible to each other. In such a case, the molecules will not be able to travel very far upon cooling, and sink marks, such as that shown in Figure 11-23, will be minimized. If the pressure (or time the pressure is applied) is too low, there

Figure 11-23. Sink marks are another form of excessive shrinkage.

will be excessive voids between the molecules, especially in areas where two walls meet. When the resin cools, these voids will collapse, bringing cooled material into them and causing excessive shrinkage and resultant sink marks. In initial molding trials, first estimate injection pressure and time, then adjust until a molded part is formed but is just short of complete filling. Increase the pressure and time in 10-percent increments until a flash-free, completely filled part is obtained. This packs out the mold and helps minimize sink marks.

Mold

Excessive rib thickness. Ribs are normally designed into a part to add strength in a given area. If the rib thickness is the same as the adjoining wall thickness, an excessively thick area is created at the junction of the rib and the wall. This thicker area takes longer to cool, and as it does, it pulls in the already cooled and solidified area around it, resulting in a sink mark. The rib wall should be designed to be no more than 60 percent of the adjoining part wall. Thus, if the part wall is nominally 0.090 in. (0.23 cm), the thickness of the rib should not exceed 0.054 in. (0.14 cm). This keeps the junction area relatively thin so it will cool at the same rate as the surrounding areas and minimize (or even eliminate) sink marks.

Material

Excessive regrind use. Regrind material is usually in the form of much larger pellets than virgin material because of the nature of the regrinding process. These larger, inconsistent particles do not nest well with the virgin particles and gaps are formed that trap air during the melting process. This trapped air impedes the ability of the molten plastic to be packed into the mold. Where trapped air bubbles exist, sink marks may form. Limit the use of regrind to 10 to 15 percent in these cases. If more regrind than that is generated, try to use it on another product, or package it up and sell it to a broker.

Operator

Early gate opening. If an operator opens the gate too soon, the cycle is shortened and the molten material may not have solidified enough to restrain still-cooling material from shrinking too much. This may cause sink marks, especially at wall junctions and around bosses. Instruct the operator on the importance of maintaining consistent cycles. Demonstrate defects caused by early gate openings. If possible, use the automatic cycle function of the molding machine to minimize operator influence.

Splay (Silver Streaking)

Machine

Barrel temperature too high. If the barrel temperature is too high, the resin will decompose and begin to char or carbonize. The charred particles will float to the resin surface during injection. The result is a spray of charred particles, on the surface of the molded part, which fans out in a direction emanating from the gate location (Figure 11-24). Decrease the barrel temperature to allow the plastic to stay molten without degrading and charring. The particles will bond together as designed and splay will be eliminated.

Figure 11-24. High material temperatures, improperly sized gates, and excessive moisture can result in splay. Cutting the barrel temperature, resizing the gates, and ensuring proper material moisture levels will eliminate this defect.

Mold

Gates too small. Gates that are too small will cause restrictive friction to the flow of the molten plastic, and this can cause degradation of the material at that spot in the mold. The degraded, decomposed material enters the cavity and may be forced to the surface in the form of a typical splay pattern. Examine the gates to make sure there are no burrs. Enlarge the gates so the depth is 50 percent of the wall thickness the gate is entering. The width can be increased until it is as much as 10 times the depth without affecting cycle times.

Material

Excessive moisture. If the material was not properly dried, the excessive moisture will turn to steam as it travels through the heating barrel. This steam becomes trapped and is carried into the mold cavity, where it is usually forced to the surface and spread across the molding surface of the cavity. It appears as streaks of silvered char, which is splay. Make sure that all materials are properly dried. Even materials that are not hygroscopic (such as nylon) must have surface moisture removed before molding. Drying conditions are critical, and material suppliers have documented conditions for specific materials and grades.

Operator

Inconsistent cycles. Erratic cycling will cause hot spots in the heating barrel. Material will become degraded in these areas and may char. These charred particles enter the melt stream and eventually the cavity where they are fanned out across the molding surface, appearing as splay. Instruct the operator on the importance of maintaining consistent cycles. Demonstrate by showing defective parts created by inconsistent cycles.

Warpage

Machine

Inadequate injection pressure or time. If too little injection pressure or time is used, the plastic material will tend to cool down and solidify before the mold is packed out. Then the individual molecules of the plastic are not packed together, leaving them space to move into as the part is cooled. They relax during the cooling period and are allowed to move about. While the outer skin of the product may be solid, the internal sections are still cooling and the movement of molecules here determines the degree of warpage. Figure 11-25 shows an example of an unacceptable degree of warpage. Increase

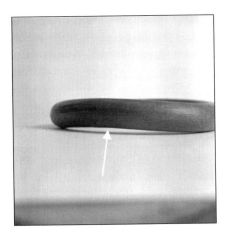

Figure 11-25. Improper flow rates and premature cool-down contribute to unacceptable warpage.

injection pressure or time to contain the cooling molecules in a rigid position (packed) until they are solid enough to prevent movement.

Mold

Mold temperature too low. Generally, a hot mold will cause the material to stay molten longer than a cold mold and allow the molecules to be packed tightly together. This results in a very dense part that minimizes the tendency for warpage. Increase the temperature to the point at which the material has the proper flow and packs the mold. Make adjustments in increments of 10° F (5.6° C) and allow 10 full shots between changes to allow the machine to stabilize.

Material

Improper flow rate. It is always best to use the stiffest flow rate possible in order to achieve the greatest property values. However, a material that is too stiff may not flow fast enough to pack the mold before it cools and solidifies. Stresses may be set up as the material stretches in an effort to fill the mold. This stretching results in warpage when the part is ejected from the mold and the stresses are slightly relieved. Use a material that has the fastest flow rate possible without causing flash.

Operator

Improper handling of parts. When the finished parts are ejected from the mold, they are still warm enough to distort if forces are applied to them. They should cool in ambient conditions for at least six full cycles before they can be packed away. An operator who is rough in handling the ejected parts can cause warpage by distorting the parts during handling. Train operators in the proper handling of still-warm molded parts.

TROUBLESHOOTING TIPS

Note that in the remedies listed, the solutions for each category tend to be repeated over and over for different defects. Several other solutions are available for specific situations, of course, but the ones listed are the most common.

- In the *machine* category, *excessive pressure* and *insufficient pressure* are responsible for the majority of the defects.
- In the *mold* category, *mold temperature* and *gate design* are responsible for the majority of the defects.
- In the *material* category, *moisture* and *improper flow rate* are responsible for the majority of the defects.
- In the *operator* category, *inconsistent cycles* are responsible for the majority of the defects.

This shows that many of the defects that occur in the molding process can usually be attributed to just one or two causes in any category. Understanding this makes troubleshooting injection molding less of a mystery and more of a science.

Successful troubleshooting usually requires making changes to an existing process. These changes will sometimes have an immediate effect, but in all cases, any changes will also have long-term effects. This is because the total molding process requires a certain amount of time to stabilize after any change is made. For instance, an increase in barrel temperature will alter the flow rate of a material after only a few minutes, but that increase also has an effect on the injection speed after a few more minutes because the material is easier to inject. A faster injection speed may initiate a tendency for flash to form after a few more minutes, and so on. Two major rules apply when making adjustments to molding parameters:

1. Make only one change at any time.
2. Allow the machine to stabilize for a period of 10 to 20 cycles after any single change is made to the process.

Troubleshooting can be a time-consuming process because of the amount of time required to allow the machine to stabilize between changes. However, without that stabilizing time, so many changes can be made that no one could determine which change actually solved the original problem. Of course the major concern is that, when many changes are made at once, the entire process quickly goes out of control, and runs in total confusion. So, a troubleshooter must be objective in analysis, be selective in solution, and most of all be patient in activity.

RULES TO MOLD BY

Here are a few rules of thumb. Some of them may seem unusual, others may seem fairly evident. All of them, however, will prove helpful. Some of them are the result of scientific research while others evolved through trial and error. Regardless of their origin, they are helpful tools for solving problems of plastic injection molding.

Back pressure: Start at 50 psi (345 kPa) and increase in 10-psi (69-kPa) increments. Do not exceed 500 psi (3447 kPa).

Barrel temperature: Set the rear zone as the coldest and the front zone as the hottest, in 10° F (5.6° C) increments.

Booster time: The time allowed for booster pressure should be adjusted in increments of 1/10 second. Allow two to three cycles between adjustments so the machine can stabilize.

Cold slug well: The diameter and depth of a cold slug well should be identical to the dimensions of the largest inside diameter of the sprue bushing.

Clamp tonnage: Clamp tonnage requirements are calculated by determining the projected area (bird's eye view) of the part to be molded and multiplying that by approximately 5 tons per square in. (68,950 kPa). An additional 10 percent should be added for each inch (2.5 cm) of depth to be molded, and a 10-percent factor should be added for safety.

Cushion: The most effective cushion is usually found at 1/8 in. (0.3 cm). Never exceed 1/4 in. (0.6 cm).

Cycle time: A part with a maximum wall thickness of 0.100 in. (0.25 cm) should have a cooling time of 15 seconds and an overall cycle time of from 20 to 25 seconds.

Cycle time adjustments: Adjust cycle times in overall increments of 5 percent. Thus, for a 30-second cycle, any single increase or decrease in cycle should total no more than 1 1/2 seconds. Allow a minimum of 10 cycles for stabilizing between each cycle adjustment.

Draft angle: The draft angle should be a minimum of 18 per side of wall.

Drying material: Always dry material for a minimum of 2 hours at the supplier's recommended temperature before using. Never use a material straight "from the bag" even if the supplier states that it would be dry enough to use that way. When using a hopper dryer, make sure that it is of sufficient size to hold the material for 2 hours before it reaches the barrel.

Gating: Always gate into a thick section before a thin section.

Gate size: Always start with a gate depth that is equal to 50 percent of the wall thickness being gated into.

Hold time (cushion time): Use holding pressure against the cushion for a minimum of 2 seconds. Increase in 1/2-second increments. The hold time need only be long enough to solidify the gate area of the part.

Injection pressure and speed: When changing injection pressure or speed, do so in increments not to exceed 10 percent of the starting value. Allow 10 full cycles between adjustments for the machine to stabilize.

Melt temperature: A change of each $10°$ F ($5.6°$ C) in the melt temperature will require 10 cycles before the barrel temperature has fully stabilized. Thus a $30°$ F ($17°$ C) increase or decrease will require a minimum of 30 cycles to stabilize.

Mold temperature: (1) A change of each $10°$ F ($5.6°$ C) in the mold temperature will require 20 cycles before the mold steel temperature has fully stabilized. Thus a $30°$ F ($17°$ C) increase or decrease will require a minimum of 60 cycles to stabilize. An aluminum mold will require only one cycle for each change. (2) A molded part will always try to stay on the hottest half of a mold. Warpage and sticking might be controlled by keeping this thought in mind. (3) A hot mold will produce a part with a finish that has more gloss than a part molded on a cold mold. A hot mold will also produce a darker part.

Nozzle temperature: In most cases, the nozzle temperature should be set to be equal to, or no more than, $10°$ F ($5.6°$ C) hotter than the front zone of the heating cylinder. Changes should be made in no more than $1°$ F ($0.56°$ C) increments.

Purging: When changing to a new material or color, approximately 20 full cycles of proper purging is required before the older material is sufficiently removed. The purging cycle should be timed the same as the normal production cycle. Do not purge one shot right after the other without the proper delay between shots.

Regrind: Up to 20 percent regrind can safely be added to virgin material, depending on the heat sensitivity of the material. Anything higher than 20 percent needs to be carefully analyzed. Of course, this assumes that the regrind is of high quality and not degraded.

Rpm of screw: Start at 100 rpm and adjust in 10-rpm increments.

Screw clearance in barrel: This clearance should never be allowed to exceed 0.005 in. (0.013 cm) per side, or 0.010 in. (0.025 cm) between the outside diameter of the screw and the inside diameter of the barrel.

Support pillars: The allowed projected molding area of a mold increases by 100 percent for each support pillar that's added.

Venting: Each vent should be a minimum of 1/8 in. (0.3 cm) wide. There should be enough venting to be equivalent to 30 percent of the measured perimeter around the parting line of the cavities.

Wall thickness: Ideally, any wall thickness should not exceed any other wall thickness by more than 25 percent. Thus, a part with a wall that is 0.100 in. (0.25 cm) thick should not have any other wall that exceeds 0.125 in. (0.31 cm) thick, or any wall less than 0.075 in. (0.19 cm) thick.

SUMMARY

Troubleshooting is a process that should be performed with objectivity and patience, and common sense should prevail. Random efforts at finding solutions will only complicate matters and create an out-of-control situation. Selected parameter changes should be performed only one at a time, and the process should be allowed to stabilize before any additional changes are made.

The primary culprit in molding is the molding machine (60 percent of the time), followed by the mold (20 percent), the raw material (10 percent), and the operator (10 percent). Unfortunately, the material and the operator are usually the first to be blamed for defective parts.

Observation is the first tool the troubleshooter should utilize. By visualizing what happens to the plastic as it travels from the hopper through the heating cylinder, and through the flow path to the cavity image, the troubleshooter can determine what may have changed to cause defects.

QUESTIONS

1. What is the primary slogan to keep in mind concerning trouble-shooting?
2. Where does one common source of troubleshooting assistance come from?
3. Not considering initial product design, what are the four root causes of most injection-molded defects?
4. What percentage of the total are each of the four causes mentioned above responsible for?
5. What type of mindset should a troubleshooter have when approaching a defect problem?
6. What three items should a troubleshooter apply to solving a defect problem?
7. What percentage (range) of the barrel capacity should be emptied every cycle?
8. How does the operator often contribute to defective molded parts?
9. What is the term used to describe a material that absorbs moisture from the atmosphere?
10. What happens to moisture in the material as it travels through the heating cylinder of the machine?

Glossary

Additive: A substance added to a plastic compound to alter its characteristics. Examples are plasticizers and flame retardants.

Alloy: A combination of two or more plastics that form a new plastic. See *blend*.

Amorphous: A plastic material in which the molecular structure is random and becomes mobile over a wide temperature range. See *crystalline*.

Anisotropic shrinkage: Shrinkage that occurs more in one direction than another (usually in the direction of flow; reinforced materials shrink more across the direction of flow).

Anneal: To heat a molded part up to a temperature just below its melting point and slowly cool to room temperature. This relieves molded stresses. See *conditioning*.

Automatic operation: The term used to define the mode in which a molding machine is operating when there is no need for an operator to start each cycle.

Barrel: A metallic cylinder in which the injection screw (or plunger) resides in the molding machine. Also called *cylinder*.

Blend: A mixture of two or more plastics.

Boss: A projection of the plastic part, normally round, which is used to strengthen an area of a part, provide a source of fastening, or provide an alignment mechanism during assembly.

Cartridge heaters: Pencil-shaped electrical heater devices sometimes placed in molds to raise the temperature level of the mold. Especially beneficial when molding high-temperature crystalline materials.

Cavity: A depression or female portion of the mold that creates the external plastic part surface.

Check ring: A ring-shaped component that slides back and forth over the tip end of the screw. The check ring eliminates the flow of molten material backward over the screw during the injection process.

Clamp force: The force, in tons or kilonewtons, that the clamp unit of a molding machine exerts to keep the mold closed during injection.

Clamp unit: The section of the molding machine containing the clamping mechanism. It is used to close the mold and keep it closed against injection pressure created by the injection process. The clamp unit also contains the ejection mechanism.

Cold slug well: A depression (normally circular) in the ejection half of an injection mold, opposite the sprue, designed to receive the first front, or "cold" portion, of molten plastic during the injection process.

Compression ratio: A factor that determines the amount of shear that is imparted to plastic material as it travels through the barrel. It is determined by dividing the depth of the screw flight in the feed section by the depth of the screw flight in the metering section.

Conditioning: Exposing a molded part to a set of conditions (such as hot oil) that impart favorable characteristics to the product. See *anneal*.

Cooling channels: Drilled holes or channels machined into various plates or components of an injection mold to provide a flow path for a cooling medium (such as water) in order to control the temperature of the mold.

Core: *(1)* An extended or male portion of the mold that creates the internal plastic part surface. *(2)* A pin or protrusion designed to produce a hole or depression in the plastic part.

Counterbore: A recessed circular area. Commonly used to fit the head of an ejector pin (return pin, sucker pin, etc.) in the ejector plate.

Crystalline: A plastic material in which the molecular structure becomes mobile only after being heated above its melting point. See *amorphous*.

Cushion: A pad of material left in the barrel at the end of the injection stroke. It is above the amount needed to fill the mold and acts as a focus point for holding pressure against the cooling melt.

Cycle: The total amount of time required for the completion of all operations needed to produce a molded part. Sometimes referred to as the *gate-to-gate* time, meaning the time from when an operator first closes the gate until the time the operator closes the gate again for starting the next cycle.

Cylinder: See *barrel*.

Decompression: A method of relieving pressure on the melt after preparing it for injection during the upcoming cycle. This minimizes the drooling that occurs when a shutoff nozzle is not utilized.

Defect: An imperfection in a molded part that results in the product not meeting original design specifications. These defects can be visual, physical, and/or hidden.

Draft: An angle (or taper) provided on the mold to facilitate removal (ejection) of the molded part.

Ejector half: The half of the mold mounted to the moving platen of the injection machine. Sometimes called the *live* half or the *movable* half because it moves. This half of the mold usually contains the ejection system.

Ejector pin: A pin, normally circular, placed in either half of the mold (usually the ejector half) which pushes the finished molded product, or runner system, out of a mold. Also referred to as a *knockout* pin, for obvious reasons.

Feed throat: The area at the rear end of the injection unit that allows fresh plastic to fall from the hopper into the heating barrel.

Feed zone: The area of the screw that is at the rear and receives fresh material from the feed throat.

Filler: Specific material added to the basic plastic resin to provide particular chemical, electrical, physical, or thermal properties.

Flash: A thin film of plastic that tends to form at parting line areas of a mold. May also be found in vent areas and around ejector pins. Flash is caused by too great a clearance between mating metal surfaces, which allows plastic material to enter.

Flight: The helical metal thread structure of the injection screw.

Gate: An opening at the entrance of a cavity (end of the runner system) that allows material to enter.

Granulator: A machine designed to grind rejected premolded plastic (products or runners). The material generated by this process is called *regrind*.

Guide pins: A pin (usually circular) that normally travels in a bushing to provide alignment of two unattached components, such as the two halves of an injection mold. Also called *leader pins*.

Heater bands: Bracelet-shaped electrical heaters placed around the outside circumference of the heating barrel.

Heating cylinder: The section of the injection-molding machine in which the plastic resin is heated to the proper molding temperature prior to injection into the mold.

Heating zone: An area of the heating barrel that is controlled by a temperature controller attached to a set of heater bands. There are four major zones: rear, center, front, and nozzle.

Hopper: A funnel-shaped container mounted over the feed throat of a molding machine. It holds fresh material to be gravity-fed into the feed zone of the heating barrel. Hoppers are normally designed to hold an average of 2 hours' worth of material for a given machine size.

Hydraulic clamp: A large hydraulic cylinder that opens and closes the clamp unit of a molding machine.

Hygroscopic: A term applied to those plastics (such as ABS and nylon) that absorb moisture from the atmosphere.

Injection capacity: A rating of the maximum amount of plastic material, in ounces or grams, that a machine can inject in a single stroke of the injection screw or plunger. It is based on the specific gravity of polystyrene as a standard.

Injection molding: The process of pushing a molten plastic material into a relatively cool mold in order to produce a finished product.

Injection pressure: The pressure that performs the initial filling of the mold. It is supplied by the injection screw or plunger as it pushes material out of the heating barrel and into the mold.

Injection unit: The section of the molding machine that contains the injection components, including the hopper, heating cylinder, screw (or plunger), nozzle, and heater bands.

Isotropic shrinkage: Shrinkage that occurs equally in all directions. See *anisotropic shrinkage*.

Knockout pin: See *ejector pin*.

L/D ratio: The result of a calculation that divides the entire length of flighted area on a screw by its nominal diameter.

Land: Describes the area in which the gate, or vent, resides. It can also be thought of as the *length* dimension in the L, W, H terminology used for describing the dimensions of the gate or vent. See also *shutoff land*.

Leader pins: See *guide pins*.

Manual operation: The term used to define the mode in which a molding machine is operating when there is a need for an operator to start and finish each phase of the total cycle.

Mechanical clamp: See *toggle clamp.*

Melt: Molten plastic prior to injection into a mold. A proper melt has the consistency of warm honey.

Metering zone: The area of the screw at the front end which contains properly melted plastic that is ready to be injected.

Mold: The entire tool (cavity, core, ejectors, etc.) needed to produce molded parts from molten plastic material.

Monomer: A molecular unit of an organic substance, usually in the form of a liquid or gas. See *polymer.*

Moving platen: The platen of a molding machine that travels (opens and closes). It is connected to the clamp unit and is the mounting location for the B, or traveling, half of the mold.

Nonreturn valve: A mechanism mounted in (or at) the nozzle of the injection machine that operates to shut off injection flow at the end of the injection cycle. It prevents material from the upcoming shot from drooling out of the nozzle when the mold opens to eject parts from the previous shot.

Nozzle: A device mounted at the end of the heating barrel that focuses plastic material to flow from the machine into the mold.

Pad: See *cushion.*

Parting line: A plane at which two halves of a mold meet. Also applies to any other plane where two moving sections come together and form a surface of a molded part.

Plastic: A complex organic compound (usually polymerized) that is capable of being shaped or formed.

Platens: The flat surfaces of a molding machine onto which the two halves of the mold are mounted. One is stationary and the other travels. There is a third platen (stationary) at the clamp end of the machine which serves as an anchoring point for the clamp unit.

Plunger: The injecting member of a nonscrew molding machine. Plungers do not rotate (auger) to bring material forward in preparation for the next cycle, nor do they blend the material as a screw does.

Polymer: A group of long chains of monomers, bonded together in a chemical reaction to form a solid. This term is often used interchangeably with *plastic,* but there can be a difference.

Purging: A process of injecting unwanted plastic material from the injection cylinder into the atmosphere for the purpose of changing materials, changing colors, or removing degraded material. Also, the name given to the mass of material that is purged.

Reciprocating screw: A helical flighted metal shaft that rotates within the heating cylinder of a molding machine, shearing, blending, and advancing the plastic material. After rotating, the screw is pushed forward to inject the plastic into the mold. Also referred to simply as the *screw*.

Regrind: Plastic material formed by granulating premolded material. Regrind is material that has been exposed to at least one heating cycle.

Residence time: The total amount of time that the plastic material *resides* in the heated barrel before being injected.

Runner: Grooves or channels cut into either or both halves of the injection mold to provide a path for the molten plastic material to be carried from the sprue to the gate(s) of the cavity.

Screw: See *reciprocating screw*.

Screw speed: The rotating speed of the screw as it augers new material toward the metering zone. It is expressed in revolutions per minute (rpm).

Secondary operation: Any activity performed after the molding process to produce a finished product suitable for its designed purpose.

Semiautomatic operation: The term used to define the mode in which a molding machine is operating when there is a need for an operator to start each cycle.

Shot: A term given to the total amount of plastic material that is injected (or shot) into a mold in a single cycle.

Shot capacity: See *injection capacity*.

Shutoff land: A raised area of the mold surface surrounding the cavity image. This area is usually between 0.002 and 0.003 in. (0.005 and 0.008 cm) high and approximately 0.5 in. (1.27 cm) wide, and is used to focus clamping pressure on the mold. The use of a shutoff land reduces the amount of force required to keep a mold closed against injection pressure.

Slide: A section of the mold that is made to travel at an angle to the normal movement of the mold. Used for providing undercuts, recesses, etc.

Sprue: The plastic material that connects the runner system to the nozzle of the heating cylinder of the molding machine. It is formed by the internal surface of a bushing that joins the mold to the machine's nozzle.

Sprue bushing: A hardened bushing that connects the mold to the molding machine nozzle and allows molten plastic to enter the runner system.

Stationary platen A: The platen at the injection end of the molding machine that does not travel. It contains the A half of the mold and locates the mold to the nozzle of the injection unit. The moving platen travels between this platen and stationary platen B.

Stationary platen B: The platen at the clamp end of the molding machine that does not travel. The moving platen travels between this platen and stationary platen A.

Stress: A resistance to deformation from an applied force. Molded plastic products tend to contain stresses molded in as a result of forces applied during the injection process. These stresses may result in fractures, cracks, and breakage if they are released during use of the product.

Suck back: See *decompression*.

Support pillar: A circular rod mold component used to support the ejector half of the mold. It is required because of the tremendous amount of pressure exerted against the B plate during the molding injection phase.

Thermocouple: A device made of two dissimilar metals which is used to measure the temperature of a heated area such as a barrel or nozzle. It sends a signal to a controller which then turns off or on to control the temperature of that area.

Thermoplastic: A plastic material that, when heated, undergoes a physical change. It can be reheated, thus reformed, repeatedly. See *thermoset*.

Thermoset: A plastic material that, when heated, undergoes a chemical change and cures. It cannot be reformed, and reheating only degrades it. See *thermoplastic*.

Tie bars: Large-diameter rods that connect *stationary platen A* to *stationary platen B*. The moving platen contains bushings that are used for sliding over the tie bars, allowing the moving platen to travel between the two stationary platens.

Toggle clamp: A mechanical scissors-action system to open and close the clamp unit of a molding machine. It is operated by a relatively small hydraulic cylinder.

Transition zone: The area in the center of the screw (between the feed zone and metering zone). This section has a tapering flight depth condition that compresses the plastic material in preparation for injection.

Undercut: A recess or extension on the molded part, located in such a way as to prevent or impede ejection of the part by normal machine operation.

Vent: A shallow groove machined into the parting line surface of a mold in order to allow air and gases to escape from the cavity, or runner, as the molten plastic is filling the mold. Sometimes also located on ejector and core pins.

Vented barrel: A heating barrel designed with an automatic venting port that allows moisture and gases to escape from molten plastic prior to injection into a mold.

Answers to
Chapter Questions

Chapter 1

1. A contest was held by a billiard ball manufacturer to find a replacement for ivory balls. This prompted John Wesley Hyatt and his brother Isaiah to injection-mold cellulose material.
2. John and Isaiah Hyatt invented the machine in 1868. The patent was issued in 1872, though, so either year should be considered correct.
3. John Hendry received a patent for the first screw-injection machine in 1946. There is no record as to when he actually began construction or made a model.
4. Any of the following: faster cycles, less energy (resources), lower costs, efficient blending or mixing (homogenizing), more consistent heating through the melt, accurate control of shot size (charge).
5. It can be utilized for creating marble finishes.
6. *Desktop Manufacturing* can be defined in this case as molding products in relatively few cavities, on high-volume equipment, small enough to fit in an area no larger than a desktop.
7. (*a*) 18,000.
8. An *alliance* might be defined as a combining of talents, resources, or expertise in an attempt to make the allying parties more efficient and productive in their respective efforts.

Chapter 2

1. Polystyrene (styrene).
2. Ideally 50 percent, but never outside the range of 20 to 80 percent.
3. The specific gravity value of the selected material is divided by the specific gravity value of polystyrene.
4. Rear, center, and front. The fourth is the nozzle zone.
5. (*b*) 2 hours.
6. 20,000 psi (138,000 kPa).
7. To keep the mold closed against injection pressure.
8. The method used for determining the required clamp force is to take the projected area of the part to be molded and multiply that number by a factor of from 2 to 8 (USCS) or 27,580 to 110,320 (SI).

9. The projected area of a part is found by multiplying the *L* dimension by the *W* dimension (length × width).
10. (*a*) - Damage may occur to the mold and to the machine.
    (*b*) - The part may flash, nonfill, or both.

Chapter 3

1. Pressure, temperature (heat), time, and distance.
2. Heater bands on the outside and shear action from the screw on the inside.
3. By using an insulation blanket on the outside of the barrel.
4. (*d*) One month.
5. Injection pressure performs the initial filling of the mold, while hold pressure maintains pressure after filling.
6. Pressure applied after injection to ensure consistency in part weight, density, and appearance.
7. Hydraulic: adjustable pressure settings; mechanical: difficult to blow open if injection pressure is too great.
8. The time from one point in a cycle (such as the safety gate closing) to the same point in the next cycle.
9. To relieve the vacuum that was created during the molding process.
10. Proper distance settings will minimize the amount of time involved for the overall cycle.
11. Material costs, labor charges, machine rate, and tooling.
12. The machine hour rate can be defined as the hourly costs involved for the operation of a machine, and includes such items as overhead, management salaries, plant maintenance, etc.
13. Charging for tooling costs by including a portion in the price of each piece produced.
14. It is a one-time charge for installing and removing the mold for a specific production run.
15. Approximately 8 percent of the original mold cost. The answer could also be $4,000 because that is the example used in the book.

Chapter 4

1. To produce the highest-quality product at the lowest possible cost.
2. Part quality requirements are usually determined through joint discussions and agreements between the owner of the product design and the manufacturer of the product.
3. (*a*) Any of the following: less shrinkage, higher gloss, less warp, harder to eject.
   (*b*) Any of the following: more shrinkage, less gloss, more warp, easier to eject.

4. One is needed for startup and another for running after machine stabilization.
5. Material gets too hot in the feed throat area, causing the pellets to stick together, blocking material from falling naturally out of the hopper into the feed throat.
6. The same as, or 10° F (5.6° C) hotter than the front zone.
7. Reduced energy costs because heat is directed against the barrel walls.
8. By creating a diameter that has the same area as the total cross-sectional area of all the runners that are fed from it.
9. Reduced material waste (less cost), and faster cycle times.
10. Stress is a resistance to deformation from an applied force.
11. 1° of draft per side.
12. To allow the molded part to be removed easily from the cavity after solidifying. It is necessary because of the vacuum that's created during molding by displacing trapped air in the cavity with plastic.
13. Because the moisture turns to steam (gas) when exposed to molding temperatures and this causes a variety of defects.
14. One that absorbs moisture directly from the surrounding atmosphere.

Chapter 5

1. The machine operator.
2. The operation of the machine gate.
3. The parts, the mold, and the machine.
4. The operator should *not* continue to cycle the machine.
5. Safety and contamination of material.
6. Input from the operator helps make the production of a molded product successful and profitable.

Chapter 6

1. Any complex, organic, polymerized compound capable of being shaped or formed.
2. Thermoplastic: a plastic material which, when heated, undergoes a *physical* change. It can be reheated, and reformed, over and over again. Thermoset: a plastic material which when heated undergoes a *chemical* change and cures. It cannot be reformed, and reheating only degrades it.
3. Amorphous (am-OR-fuss) materials are those in which the molecular structure is random and becomes mobile over a wide temperature range.

4. Crystalline (CRISS-tull-in) materials, on the other hand, are those in which the molecular structure is well-ordered and becomes mobile only after being heated to its melting point.

5. Any of those listed in the chapter. ABS and acrylic are amorphous, for example, while acetal and cellulose butyrate are crystalline.

6. Carbon and hydrogen.

7. A suggested definition of polymerization: a reaction caused by combining monomers with a catalyst, under pressure, and with heat.

8. Heat starts the molecules moving.

9. Pressure aligns the molecules and pushes them along the flow path into the mold.

10. Cooling halts the molecular action.

11. Processibility goes down.

12. Fillers are used to change the properties of a resin.

13. Reinforcements are used to impart strength to a resin.

14. ASTM test D-1238 determines the melt flow index (flowability) of a plastic material.

Chapter 7

1. Because the plastic molded product is formed within the mold.

2. The purpose of the sprue bushing is to seal tightly against the nozzle of the injection barrel of the molding machine and to allow molten plastic to flow from the barrel into the mold.

3. Flash is material that seeps from a mold. Causes are too little clamp pressure, injection pressure that is too great, and a mold that is wearing.

4. Reduced material waste and faster cycles.

5. Body, head, and face.

6. By a knockout rod mounted to the machine.

7. The gate should be located in the thickest section of the part.

8. A circular (round) runner cross section is preferred.

9. Air is trapped in a mold because when the mold closes it forms a tight parting line seal around air in the machined cavity image and runner.

10. Vents can be added.

11. Because if the air is allowed to escape from the runner, it won't enter the cavity from which it would have to be vented.

Chapter 8

1. The three main types of units for drying material are ovens, hopper dryers, and floor dryers. (Vented barrel is not a main type.)

2. The TVI (Tomasetti volatile indicator) test.
3. Mechanical, vacuum, and positive pressure.
4. 50 ft (15 m).
5. Mixing regrind (or color) with virgin material.
6. Grinding up scrapped parts and runner systems.
7. Two. One for each half.
8. Water and oil (mineral or silicone).
9. Flexible and rigid.
10. Flexible.

Chapter 9

1. Any operation performed on a product after it is molded.
2. Any two of the following: when volumes are small; when tooling costs are excessive; when the time to build a mold will jeopardize schedules; when a labor-heavy environment exists.
3. 20 to 40 kHz (20,000 to 40,000 cycles per second).
4. Any two of the following: polymer structure, melt temperature, stiffness, moisture content, flow rate, mold release agents, plasticizers, flame retardants, regrind, colorants, resin grades, fillers, reinforcements.
5. Nitrile elastomer.
6. 1000 to 5000 rpm.
7. To ensure adequate bonding of the decorative material to the plastic material.
8. Any two of the following: detergent wash, flame treatment, corona discharge, plasma process, acid wash.
9. Applied finishes are utilized after the product has been removed from the mold, while in-process finishes are utilized while the part is being molded.
10. Applied includes: painting, plating, vacuum metallizing, hot stamping, pad printing, and screen printing. In-process includes: molded-in color, molded-in symbols, two-color (two-shot) molding, textured surfaces, and in-mold overlays.

Chapter 10

1. ASTM and ISO.
2. Improper test results.
3. Any three of the following: dielectric strength, dielectric constant, volume resistivity, surface resistivity, arc resistance.
4. Any three of the following: shrinkage rate, density, water absorption, moisture content, melt flow index.

5. Any three of the following: tensile strength, compressive strength, flexural strength, creep, impact resistance.
6. Any three of the following: melt point, heat deflection temperature, Vicat softening temperature, flammability, limiting oxygen index.
7. Failure analysis is performed on a product that fails after it has been molded. Troubleshooting is performed on the molding process as it takes place or shortly after.
8. A resistance to deformation from an applied force.
9. Differential Scanning Calorimeter.
10. Two peaks are generated on a curve, and the areas under the peaks are compared to calculate a ratio equaling a percentage of regrind to virgin.
11. A peak from a known, fully crystallized sample is compared to the peak of the unknown sample, and the areas under the peaks are compared to calculate a ratio of the crystallinity of each.
12. The furnace method and the TGA method. The torch method is also used.

Chapter 11
1. Keep it simple.
2. The material suppliers.
3. Machine, mold, material, and operator.
4. Machine = 60 percent, mold = 20 percent, material = 10 percent, and operator = 10 percent.
5. An objective mind.
6. Objectivity, simple analysis, and common sense.
7. Between 20 and 80 percent.
8. By running inconsistent cycles.
9. It is said to be *hygroscopic*.
10. It turns to steam (gas).

Index

medium, 85-87
pin(s), 86, 87
postmold, 111
time, 47, 77, 85, 88
timer (dwell timer), 47, 48
water, 226
core pins, 140, 237
corners
radiused, 116
squared, 117
corona discharge process, 183
cushion (pad), 38, 46, 52, 98
distance, 77, 98
holding, 53, 99
cycle(s), 2, 11, 14, 31, 121
consistent, 222
gate-to-gate, 62
inconsistent, 92, 222, 224, 227-230,
232, 234, 237, 239, 240, 242-244,
248, 249
time, 21, 29, 44, 49-51, 62, 76, 77, 88,
94, 96, 110, 150, 165, 238

D

debugging, 221
decorating procedures, 182
postmold, 182
defect(s), 219-221, 234, 244-246
black specks or streaks, 221
blistering, 222, 223
blush, 192, 223, 224, 231
bowing, 224
brittleness, 214, 225, 226
bubbles, 227
burn marks, 228
causes, 220
clear spots, 229, 230
cloudiness, 231
cracking, 52, 112, 233, 234
crazing, 112, 184, 234
delamination, 234-236
discoloration, 110, 235, 237
flash(ing), 24, 42, 50, 70, 97, 124-126,
140, 142, 235, 237, 238, 249
flow lines, 238-240
gloss (low), 239

jetting, 241, 242
knit lines (weld lines), 192, 242
nonfill (short shots), 238, 243, 244
and remedies, 221
shrinkage, 35, 52, 69, 105, 107, 108,
225, 244, 245
sink marks, 227, 245, 246
splay (silver streaks), 100, 118, 192,
224, 231, 247
warpage, 35, 52, 85, 112, 223, 248, 249
deformation, 112
degradation, 18, 69, 83, 92, 171, 214, 222,
235, 247
desiccant(s), 151
banks, 152
bed, 154
desktop manufacturing, 4, 8
dew-point
measurement, 155, 163
meter, 155
temperature, 155
dielectric
constant, 196, 197
strength, 196
Differential Scanning Calorimeter, 207,
212, 213
testing, 212
distance, 30, 38, 50, 65, 69
cushion, 77
ejection, 55
initial injection, 51
injection, 52
injection hold, 51, 52
mold-close, 50
mold-open, 53
screw return, 54
draft, 233
allowance, 114
angle(s), 113-116
drooling, 101
dryer units, 151, 154, 163
floor, 151-153, 163
hopper, 151, 152, 163
oven, 151, 153, 154, 163
vented barrels, 151
drying materials, 118

U

V